THE EVOLUTION
OF CAPITALISM

THE EVOLUTION
OF CAPITALISM

Advisory Editor
LEONARD SILK
Editorial Board,
The New York Times

Research Associate
MARK SILK

CAPITAL AND LABOUR;

INCLUDING

THE RESULTS OF MACHINERY

BY CHARLES KNIGHT

ARNO PRESS

A NEW YORK TIMES COMPANY

New York • 1972

Reprint Edition 1972 by Arno Press Inc.

The Evolution of Capitalism
ISBN for complete set: 0-405-04110-1
See last pages of this volume for titles.

Manufactured in the United States of America

- - - - - - - - - - - -

Library of Congress Cataloging in Publication Data

Knight, Charles, 1791-1873.
   Capital and labour.

   (The Evolution of capitalism)
   Reprint of the 1845 ed.
   1. Labor and laboring classes. 2. Machinery in
industry. 3. Labor and laboring classes--Gt. Brit.
I. Title. II. Series.
HD8389.K5 1972   331.1'1'0942   76-38272
ISBN 0-405-04125-X

CLASS I.

COPYRIGHT MISCELLANIES.

CAPITAL AND LABOUR;

INCLUDING

THE RESULTS OF MACHINERY.

BY

CHARLES KNIGHT.

KNIGHT'S WEEKLY VOLUME FOR ALL READERS

LXVII.

LONDON:

CHARLES KNIGHT & CO., LUDGATE STREET.

W. CLOWES AND SONS, STAMFORD STREET.

CAPITAL AND LABOUR;

INCLUDING

THE RESULTS OF MACHINERY.

BY CHARLES KNIGHT.

LONDON:

CHARLES KNIGHT & Co., LUDGATE STREET.

1845.

LONDON:
PRINTED BY WILLIAM CLOWES AND SONS,
STAMFORD STREET.

ADVERTISEMENT.

"THE Results of Machinery" was written by me at a period of great national alarm, when a blind rage against a power supposed to interfere with the claims of Labour was generally prevalent, and led, in the Southern agricultural districts especially, to many acts of daring violence. That little book had a most extensive sale, and is still in constant demand. Fifty thousand copies have been sold since its first publication. Within a year after that publication I wrote a second tract, "Capital and Labour," which was to form part of a Series entitled "The Rights of Industry." This Series I never could find leisure to proceed with. It has appeared to me that the two parts might be advantageously incorporated. Machinery, in connexion with Capital and Labour, is one of the great instruments of Production. In this Volume, then,

thus remodelled, the general subject of THE PRO-
DUCTION OF WEALTH is fully, though popularly
expounded.   The original Tracts were especially
addressed to *Working Men*.   This Volume is addressed to *all*.   The Statistical details are brought
up to the present time.

CHARLES KNIGHT.

September 20, 1845.

CAPITAL AND LABOUR.

CHAPTER I.

Let us suppose a man brought up in civilized life, cast upon a desert land—without food, without clothes, without fire, without tools. We see the human being in the very lowest state of helplessness. Most of the knowledge he had acquired would be worse than useless; for it would not be applicable in any way to his new position. Let the land upon which he is thrown produce spontaneous fruits—let it be free from ferocious animals—let the climate be most genial—still the man would be exceedingly powerless and wretched. The first condition of his lot, to enable him to maintain existence at all, would be that he should labour. He must labour to gather the berries from the trees—he must labour to obtain water from the rivulets—he must labour to form a garment of leaves, or of some equally accessible material, to shield his body from the sun—he must labour to render some cave or hollow tree a secure place of shelter from the dews of night. There would be no intermission of the labour necessary to provide a supply of food from hand to mouth, even in the season when wild fruits were abundant. If this labour, in the most favourable season, were interrupted for a single day, or at most for two or three days, by sickness, he would in all probability perish. But, when the autumn was past, and the wild fruits were gone, he must prolong existence like some savage tribes are reported to do—by raw fish and undressed roots.

B

The labour of procuring these would be infinitely greater than that of climbing trees for fruit. To catch fish without nets, and scratch up roots with naked hands, is indeed painful toil. The helplessness of this man's condition would principally be the effect of one circumstance ;—he would possess no accumulation of former labour by which his present labour might be profitably directed. *The power of labour would in his case be in its least productive state.* He would partly justify the assertion that man has the feeblest natural means of any animal ;—because he would be utterly unpossessed of those means which the reason of man has accumulated around every individual in the social state.

We asked the reader to *suppose* a civilized man in the very lowest state in which the power of labour may be exercised, because there is no record of any civilized man being for any length of time in such a state.

Ross Cox, a Hudson's Bay trader, whose adventures were given to the world about twelve years ago, was in this state for a fortnight; and his sufferings may furnish some idea of the greater miseries of a continuance in such a powerless condition. Having fallen asleep in the woods of the north-west of America, which he had been traversing with a large party, he missed the traces of his companions. The weather being very hot, he had left nearly all his clothes with his horse when he rambled from his friends. He had nothing to defend himself against the wolves and serpents but a stick ; he had nothing of which to make his bed but long grass and rushes ; he had nothing to eat but hips and wild cherries. The man would doubtless have perished, unless he had met with some Indians, who knew better how to avail themselves of the spontaneous productions around them. But this is not an instance of the continuance of Labour in the lowest state of its power.

The few individuals, also, who have been found exposed in forests, such as Peter the Wild Boy, and the Savage of Aveyron,—who were discovered, the one about a century ago, in Germany, the other about forty years since, in France,—differed from the civilized man cast naked upon a desert shore in this particular; their *wants* were of the lowest nature. They were not raised above the desires of the most brutish animals. They supplied those desires after the fashion of brutes. Peter was enticed from the woods by the sight of two apples, which the man who found him displayed. He did not like bread, but he eagerly peeled green sticks, and chewed the rind. He had, doubtless, subsisted in this way in the woods. He would not, at first, wear shoes, and delighted to throw the hat which was given him into the river. He was brought to England, and lived many years with a farmer in Hertfordshire. During the Scotch Rebellion, in 1745, he wandered into Norfolk; and having been apprehended as a suspicious character, was sent to prison. The gaol was on fire; and Peter was found in a corner, enjoying the warmth of the flames without any fear. The Savage of Aveyron, in the same manner, had the lowest desires and the feeblest powers. He could use his hands, for instance, for no other purpose than to seize upon an object; and his sense of touch was so defective, that he could not distinguish a raised surface, such as a carving, from a painting. This circumstance of the low physical and intellectual powers of these unfortunate persons prevents us exhibiting them as examples of the state which we asked the reader to suppose.

Let us advance another step in our view of the power of Labour. Let us take a man in one respect in the same condition that we supposed—left upon a desert land, without any direct social aid; but with some help to his labour by a small Accumulation of former industry. We have instances on record of this next state.

In the year 1681 a Moskito Indian was left by accident on the island of Juan Fernandez, in the Pacific Ocean; the English ship, in which he was a sailor, having been chased off the coast by some hostile Spanish vessels. Captain Dampier describes this man's condition in the following words :—

" This Indian lived here alone above three years; and although he was several times sought after by the Spaniards, who knew he was left on the island, yet they could never find him.   He was in the woods hunting for goats, when Captain Watlin drew off his men, and the ship was under sail before he came back to shore. He had with him his gun, and a knife, with a small horn of powder, and a few shot; which being spent, he contrived a way, by notching his knife, to saw the barrel of his gun into small pieces, wherewith he made harpoons, lances, hooks, and a long knife; heating the pieces first in the fire, which he struck with his gun-flint, and a piece of the barrel of his gun, which he hardened; having learnt to do that among the English.   The hot pieces of iron he would hammer out, and bend as he pleased with stones, and saw them with his jagged knife, or grind them to an edge by long labour, and harden them to a good temper as there was occasion.*   With such instruments as he made in that manner, he got such provisions as the island afforded, either goats or fish. He told us that at first he was forced to eat seal, which is very ordinary meat, before he had made hooks; but afterwards he never killed any seals but to make lines, cutting their skins into thongs.   He had a little house, or hut, half a mile from the sea, which was lined with goat's skin; his couch, or barbecu of sticks, lying along about two feet distance from the ground, was spread with the same, and was all his bedding.   He had no clothes left; having worn out those he brought from Watlin's ship, but only a skin about his waist.   He saw our ship the day before we came to an anchor, and did believe we were English; and therefore killed three goats in the morning, before we came to an anchor, and drest them with cabbage, to treat us when we came ashore."

Here, indeed, is a material alteration in the wealth of a man left on an uninhabited island.   He had a regular

---

* It is difficult to understand how the Indian could convert the *iron* gun-barrel into steel, which it appears, from Dampier's account, that he did.   Steel is produced by a scientific admixture of carbon with iron.   But we assume that the statement is correct, and that a conversion, partial doubtless, of iron into steel did take place.

supply of goats and fish; he had the means of cooking this food; he had a house lined with goats' skins, and bedding of the same; his body was clothed with skins; he had provisions in abundance to offer, properly cooked, when his old companions came to him after three years' absence. What gave him this power to labour profitably? —to maintain existence in tolerable comfort? Simply, the gun, the knife, and the flint, which he accidentally had with him when the ship sailed away. The flint and the bit of steel which he hardened out of the gun-barrel, gave him the means of procuring fire; the gun became the material for making harpoons, lances, and hooks, with which he could obtain fish and flesh. Till he had these tools, he was compelled to eat seal's flesh. The instant he possessed the tools, he could make a selection of what was most agreeable to his taste. It is almost impossible to imagine a human being with less accumulation about him. His small stock of powder and shot was soon spent, and he had only an iron gun-barrel and a knife left, with the means of changing the form of the gun-barrel by fire. Yet this single accumulation enabled him to direct his labour, as all labour is directed even in its highest employment, to the change of form and change of place of the natural supplies by which he was surrounded. He created nothing; he only gave his natural supplies a value by his labour. Until he laboured, the things about him had no value, as far as he was concerned; when he did obtain them by labour, they instantly acquired a value. He brought the wild goat from the mountain to his hut in the valley—he changed its place; he converted its flesh into cooked food, and its skin into a lining for his bed—he changed its form. Change of form and change of place are the beginning and end of all human labour; and the Moskito Indian only employed the same principle for the supply of his wants, which directs the labour of all the producers of civilized life into the channels of manufactures or commerce.

But the Moskito Indian, far removed as his situation was above the condition of the man without any accumulation of former labour—that is, of the man without any capital about him—was only *in the second stage in which the power of labour can be exercised,* and in which it is comparatively still weak and powerless. He laboured— he laboured with accumulation—but he laboured without that other power which gives the last and highest direction to profitable labour.

Let us state all the conditions necessary for the production of Utility, or of what is essential to the support, comfort, and pleasure of human life :—

1. *That there shall be Labour.*

The man thrown upon a desert island without accumulation,—the half-idiot boy who wandered into the German forests at so early an age that he forgot all the usages of mankind,—were each compelled to labour, and to labour unceasingly, to maintain existence. Even with an unbounded command of the spontaneous productions of nature, this condition is absolute. It applies to the inferior animals as well as to man. The bee wanders from flower to flower, but it is to labour for the honey. The sloth hangs upon the branches of a tree, but he labours till he has devoured all the leaves, and then climbs another tree. The condition of the support of animation is labour ; and if the labour of all animals were miraculously suspended for a season, very short as compared with the duration of individual life, the reign of animated nature upon this globe would be at an end.

The second condition in the production of utility is,—

2. *That there shall be accumulation of former labour, or, Capital.*

Without accumulation, as we have seen, the condition of man is the lowest in the scale of animal existence. The reason is obvious. Man requires some accumulation to aid his natural powers of labouring ; for he is not provided with instruments of labour to anything like the perfection in which they exist amongst the inferior animals. He wants the gnawing teeth, the tearing claws, the sharp bills, the solid mandibles that enable quadrupeds, and birds, and insects to secure their food, and to provide shelter in so many ingenious ways, each leading us to admire and reverence the directing Providence which presides over such manifold contrivances. He must, therefore, to work profitably, accumulate instruments of work. But he must do more, even in the unsocial state, where he is at perfect liberty to direct his industry as he pleases uncontrolled by the rights of other men. He must accumulate stores of covering and of shelter. He must have a hut, and a bed of skins, which are all accumulations, or capital. He must, further, have a stock of food, which stock, being the most essential for human wants, is called *provisions*, or things provided. He would require this provision against the accidents which may occur to his own health, and the obstacles of weather, which may prevent him from fishing or hunting. The lowest savages have some stores. Many of the inferior animals display an equal care to provide for the exigencies of the future. But still, all such labour is extremely limited. When a man is occupied only in providing immediately for his own wants—doing everything for himself, consuming nothing but what he produces himself,—his labour must have a very narrow range. The supply of the lowest necessities of our nature can only be attended to, and these must be very ill supplied. The Moskito Indian had fish, and goats' flesh, and a rude hut, and a girdle of skins; and his power of obtaining this wealth was insured to him by the absence of other individuals who would have been his competitors for what the island spontaneously produced. Had other Indians landed in numbers on the island, and had each set about procuring everything for himself, as the active Moskito did, they would have soon approached the point of starvation ; and then each would have begun to plunder from the other, unless they had found out the principle that would have given them all plenty. There wanted, then, another power to give the labour of the Indian a profitable direction, besides that of accumulation. It is a power which can only exist where man is social, as it is his nature to be ;—and where the principles of civilization are in a certain degree developed. It is, indeed, the beginning and the end of all civilization. It is itself civilization, partial or complete. It is the last and the most important condition in the production of useful commodities,—

3. *That there shall be Exchanges.*

There can be no exchanges without accumulation—there can be no accumulation without labour. Exchange is that step beyond the constant labour and the partial accumulation of the lower animals, which makes man the lord of the world.

CHAPTER II.

SOCIETY, both in its rudest form and in its most refined and complicated relations, is nothing but a system of Exchanges. An exchange is a transaction in which both the parties who make the exchange are benefited ;—and, consequently, society is a state presenting an uninterrupted succession of advantages for all its members. Every time that we make a free exchange we have a greater desire for the thing which we receive than for the thing which we give ;—and the person with whom we make the exchange has a greater desire for that which we offer him than for that which he offers us. When one gives his labour for wages, it is because he has a higher estimation of the wages, than of the profitless ease and freedom of remaining unemployed ;—and, on the contrary, the employer who purchases his labour, feels that he shall be more benefited by the results of that labour, than by retaining the capital which he exchanges for it. In a simple state of society, when one man exchanges a measure of wheat for the measure of wine which another man possesses, it is evident that the one has got a greater store of wheat than he desires to consume himself, and that the other, in the same way, has got a greater store of wine ;—the one exchanges something to eat for something to drink, and the other something to drink for something to eat. In a refined state of society, when money represents the value of the exchanges, the exchange between the abundance beyond the wants of the possessor of one commodity, and of another, is just as real as the barter of wheat for wine. The only difference is, that the exchange is not so direct, although It is incomparably more rapid. But, however the system of exchange be carried on,—whether the value of the things exchanged be determined by barter or by a price in money,—all the exchangers are benefited, because all obtain what they want, through the store which they possess of what they do not want.

It has been well said, that " Man might be defined to be an animal that makes exchanges."* There are other animals, indeed, such as bees and ants amongst insects, and beavers amongst quadrupeds, who to a certain extent are social ; that is, they concur together in the execution of a common work for a common good : but as to their individual possessions, each labours to obtain what it desires from sources accessible to all, or plunders the stores of others. Not one insect or quadruped, however wonderful may be its approaches to rationality, has the least idea of making a formal exchange with another. The modes by which the inferior animals communicate their thoughts are probably not sufficiently determinate to allow of any such agreement. The very foundation of that agreement is a complicated principle, which man alone can understand. It is the Security of individual Property. Immediately that this principle is established, labour begins to work profitably, for it works with exchange. If the principle of appropriation were not acted upon at all, there could be no exchange, and consequently no production. The scanty bounty of nature might be scrambled for by a few miserable individuals—and the strongest would obtain the best share ; but this insecurity would necessarily destroy all accumulation. Each would of course live from hand to mouth, when the means of living were constantly exposed to the violence of the more powerful. This is the state of the lowest savages, and as it is an extreme state it is a rare one,—no security, no exchange, no capital, no labour, no production. Let us apply the principle to an individual case.

The poet who has attempted to describe the feelings of a man suddenly cut off from human society, in ' Verses

* Dr. Whately's Lectures on Political Economy.

supposed to be written by Alexander Selkirk during his solitary abode in the island of Juan Fernandez,' represents him as saying, " I am monarch of all I survey."* Alexander Selkirk was left upon the same island as the Moskito Indian ; and his adventures there have formed the groundwork of the beautiful romance of ' Robinson Crusoe.' The meaning of the poet is, that the unsocial man had the same right over all the natural productive powers of the country in which he had taken up his abode, as we each have over light and air. He was alone ; and therefore he exercised an absolute, although a barren sovereignty, over the wild animals by which he was surrounded—over the land and over the water. He was, in truth, the one proprietor—the one capitalist, and the one labourer—of the whole island. His absolute property in the soil, and his perfect freedom of action, were both dependent upon one condition—that he should remain alone. If the Moskito Indian, for instance, had remained in the island, Selkirk's entire sovereignty must have been instantly at an end. Some more definite principle of appropriation must have been established, which would have given to Selkirk, as well as to the Moskito Indian, the right to appropriate distinct parts of the island each to his particular use. Selkirk, for example, might have agreed to remain on the eastern coast, while the Indian might have established himself on the western ;—and then the fruits, the goats, and the fish of the eastern part would have been appropriated to Selkirk, as distinctly as the clothes, the musket, the iron· pot, the can, the hatchet, the knife, the mathematical instruments, and the Bible, which he brought on shore.† If the Indian's territory had produced something which Selkirk had not, and if Selkirk's land had also something which the Indian's had not, they might have become exchangers. They would have passed into that condition naturally enough ;—imperfectly perhaps, but still as easily

---

* Cowper's Miscellaneous Poems.
† These circumstances are recorded in Captain Woodes Rogers' Cruizing Voyage round the World, 1712.

as any barbarous people who do not cultivate the earth, but exchange her spontaneous products.

The poet goes on to make the solitary man say, " My right there is none to dispute." The condition of Alexander Selkirk was unquestionably one of absolute liberty. His rights were not measured by his duties. He had all rights and no duties. Many writers on the origin of society have held that man, upon entering into union with his fellow-men, and submitting, as a necessary consequence of this union, to the restraints of law and government, sacrifices a portion of his liberty, or natural power, for the security of that power which remains to him. No such agreement amongst mankind could ever have possibly taken place ; for man is by his nature, and without any agreement, a social being. He is a being whose rights are balanced by the uncontrollable force of their relation to the rights of others. The succour which the infant man requires from its parents, to an extent, and for a duration, so much exceeding that required for the nurture of other creatures, is the natural beginning of the social state, established insensibly and by degrees. The liberty which the social man is thus compelled by the force of circumstances to renounce, amounts only to a restraint upon his brute power of doing injury to his fellow-men : and for this sacrifice, in itself the cause of the highest individual and therefore general good, he obtains that dominion over every other being, and that control over the productive forces of nature, which alone can render him the monarch of all he surveys. The poor sailor, who for four years was cut off from human aid, and left alone to struggle for the means of supporting existence, was an exception, and a very rare one, to the condition of our species all over the world. His absolute rights placed him in the condition of uncontrolled feebleness ; if he had become social, he would have put on the regulated strength of rights balanced by duties.

Alexander Selkirk was originally left upon the uninhabited island of Juan Fernandez at his own urgent desire. He was unhappy on board his ship, in consequence of disputes with his captain ; and he resolved to rush into a state which might probably have separated him for ever from the rest of mankind. In the belief that he should be so separated, he devoted all his labour and all his ingenuity to the satisfaction of his own wants alone. By continual exercise, he was enabled to run down the wild goat upon the mountains; and by persevering search, he knew where to find the native roots that would render his goat's flesh palatable. He never thought, however, of providing any store beyond the supply of his own personal necessities. He had no motive for that thought; because there was no human being within his reach with whom he might exchange that store for other stores. The very instant, however, that the English ships, which finally gave him back to society, touched upon his shores,—before he communicated by speech with any of his fellow-men, or was discovered by them,— he became social. He saw that he must be an exchanger. Before the boat's crew landed he had killed several goats, and prepared a meal for his expected guests. He knew that he possessed a commodity which they did not possess. He had fresh meat, whilst they had only salt. Of course what he had to offer was acceptable to the sailors ; and he received in exchange protection, and a place amongst them. He renounced his sovereignty, and became once more a subject. It was better for him, he thought, to be surrounded with the regulated power of civilization, than to wield at his own will the uncertain strength of solitary uncivilization. But, had he chosen to remain upon his island, as in after-years he regretted he had not done, although a solitary man, he would not have been altogether cut off from the hopes and the duties of the social state. If he had chosen to remain after that visit from his fellow-men, he would have said to them, before they had left him once more alone, " I have hunted for you my goats, I have dug for you my roots, I have shown you the fountains which issue out of my rocks ;—these are the resources of my dominion : give me in exchange for them a fresh supply of gunpowder and shot, some of your clothes, some of the means of repairing these clothes, some of your tools and implements of cookery, and more of your books to divert my solitary hours." Having enjoyed the benefits which he had bestowed, they would, as just men, have paid the debt which they had incurred, and the exchange would have been completed. Immediately that they had quitted his shores, Selkirk would have looked at his resources with a new eye. His hut was rudely fashioned and wretchedly furnished. He had fashioned and furnished it as well as he could by his own labour, working upon his own materials. The visit which he had received from his fellow-men after he had abandoned every hope of again looking upon their faces, would have led him to think that other ships would come, with whom he might make other exchanges, —new clothes, new tools, new materials, received as the price of his own accumulations. To make the best of his circumstances when that day should arrive, he must redouble his efforts to increase his stock of commodities,— some for himself, and some to exchange for other commodities, if the opportunity for exchange should ever come. He must therefore transplant his vegetables, so as to be within instant reach when they should be wanted. He must render his goats domestic, instead of chasing them upon the hills. He must go forward from the hunting state, into the pastoral and agricultural.

All this particular course of improvement, however, supposes that Selkirk should continue in his state of exclusive proprietor—that there should be none to dispute his right. If other ships had come to his shores—if they had trafficked with him from time to time—exchanged clothes and household conveniences, and implements of cultivation, for his goats' flesh and roots—it is probable that other sailors would in time have desired to partake his plenty ;—that a colony would have been founded—that the island would have become populous. It is perfectly clear that, whether for exchange amongst themselves, or for exchange with others, the members of this colony could not have stirred a step in the cultivation of the land without appropriating its produce ;—and they could not have appropriated its produce without appropriating the land itself. Cultivation of the land for a common stock would have gone to the establishment precisely of the same principle;—they would still have been exchangers amongst themselves, and the partnership would not have lasted a day, unless each man's share of what the partnership produced had been rendered perfectly secure to him. Without security they could not have accumulated—without accumulations they could not have exchanged—without exchanges they could not have carried forward their labours with any compensating productiveness.

Imperfect appropriation — that is, an appropriation which respects personal wealth, such as the tools and conveniences of an individual, and even secures to him the fruits of the earth when he has gathered them, but which has not reached the last step of a division of land —imperfect appropriation such as this raises up the same invincible obstacles to the production of utility; because, with this original defect, there must necessarily be unprofitable labour, small accumulation, limited exchange. Let us exemplify this by another individual case.

We have seen, in the instances of the Moskito Indian and of Selkirk, how little a solitary man can do for himself, although he may have the most unbounded command of natural supplies—although not an atom of those natural supplies, whether produced by the earth or the water, is appropriated by others—when, in fact, he is monarch of all he surveys. Let us trace the course of another man, advanced in the ability to subdue all things to his use by association with his fellow-men; but carrying on that association in the rude and unproductive relations of savage life;—not desiring to "replenish the earth" by cultivation, but seeking only to appropriate the means of existence which it has spontaneously produced;—labouring, indeed, and exchanging, but not labouring and exchanging in a way that will permit the accumulation of wealth, and therefore remaining poor and miserable. We are not about to draw any fanciful picture, but merely to select some facts from a real narrative.

CHAPTER III.

In the year 1828, there came to New York, in the United States, a white man named John Tanner, who had been thirty years a captive amongst the Indians in the interior of North America. He was carried off by a band of these people when he was a little boy, from a settlement on the Ohio river, which was occupied by his father, who was a clergyman. The boy was brought up in all the rude habits of the Indians, and became inured to the abiding miseries and uncertain pleasures of their wandering life. He grew in time to be a most skilful huntsman, and carried on large dealings with the agents of the Hudson's Bay Company, in the skins of beavers and other animals, which he and his associates had shot or entrapped. The history of this man was altogether so curious, that he was induced to furnish the materials for a complete narrative of his adventures; and, accordingly, a book, fully descriptive of them, was prepared for the press by Dr. Edwin James, and printed at New York in 1830. It is of course not within the intent of our little work to furnish any regular abridgment of John Tanner's story; but it is our wish to direct attention to some few particulars, which appear to us strikingly to illustrate some of the positions which we desire to enforce by thus exhibiting their practical operation.

The country in which this man lived so many years is that immense territory belonging to the United States, which is still covered by boundless forests which the progress of civilization has not yet cleared away. In this region a number of scattered Indian tribes maintain a precarious existence by hunting the moose-deer and the buffalo for their supply of food, and by entrapping the foxes and martens of the woods and the beavers of the lakes, whose skins they generally exchange with the white traders of Europe for articles of urgent necessity, such as ammunition and guns, traps, axes, and woollen blankets; but too often for ardent spirits, equally the curse of savage and of civilized life. The contact of savage man with the outskirts of civilization perhaps afflicts him with the vices of both states. But the principle of exchange, imperfectly and irregularly as it operates amongst the Indians, furnishes some excitement to their ingenuity and their industry. Habits of providence are to a certain degree created; it becomes necessary to accumulate some capital of the commodities which can be rendered valuable by their own labour, to exchange for commodities which their own labour, without exchange, is utterly unable to procure. The principle of exchange, too, being recognised amongst them in their dealings with foreigners, the security of property—without which, as we have shown, that principle cannot exist at all—is one of the great rules of life amongst themselves. But still these poor Indians, from the mode which they propose to themselves for the attainment of property, which consists only in securing what nature has produced, without directing the course of her productions, must be very far removed from the regular attainment of those blessings which civilized society alone offers. We shall exemplify these statements by a few details.

The country over which these people range occupies a surface that may be roughly described as five or six times as large as all England. They have the unbounded command of all the natural resources of that country; and yet their entire numbers do not exceed a few hundred thousands—that is, they do not equal the population of a moderately sized English county. It may be fairly said that each Indian requires a thousand acres for his maintenance. The supplies of food are so scanty—a scantiness which would at once cease to exist were there any cultivation— that if a large number of these Indians assembled together to co-operate in their hunting expeditions, they are very soon dispersed by the urgent desire of satisfying hunger. Tanner says, "We all went to hunt beavers in concert. In hunts of this kind the proceeds are sometimes equally divided ; but in this instance every man retained what he had killed. In three days I collected as many skins as I could carry. But in these distant and hasty hunts little meat could be brought in ; and the whole band was soon suffering with hunger. Many of the hunters, and I among others, for want of food became extremely weak, and unable to hunt far from home." In another place he says, "I began to be dissatisfied at remaining with large bands of Indians, as it was usual for them, after having remained a short time in a place, to suffer from hunger." These sufferings were not, in many cases, of short duration, or of trifling intensity. Tanner describes one instance of famine in the following words :—" The Indians gathered around, one after another, until we became a considerable band, and then we began to suffer from hunger. The weather was very severe, and our suffering increased. A young woman was the first to die of hunger. Soon after this, a young man, her brother, was taken with that kind of delirium or madness which precedes death in such as die of starvation. In this condition he had left the lodge of his debilitated and desponding parents ; and when, at a late hour in the evening, I returned from my hunt, they could not tell what had become of him. I left the camp about the middle of the night, and following his track, I found him at some distance, lying dead in the snow."

This worst species of suffering equally exists at particular periods, whether food be sought for by large or by small parties, by bands or by individuals. Tanner was travelling with the family of the woman who had adopted him. He says, "We had now a short season of plenty ; but soon became hungry again. It often happened that for two or three days we had nothing to eat ; then a rabbit or two, or a bird, would afford us a prospect of protracting the sufferings of hunger a few days longer." Again he says, "Having subsisted for some time almost entirely on the inner bark of trees, and particularly of a climbing vine found there, our strength was much reduced."

The misery which is thus so strikingly described, proceeds from the circumstance that the labour of the Indians does not take a profitable direction: and that this waste of labour (for unprofitable applications of labour are the greatest of all wastes) arises from the one fact, that in certain particulars these Indians labour without appropriation. They depend upon the chance productions of nature, without compelling her to produce; and they do not compel her to produce, because there is no appropriation of the soil, the most efficient natural instrument of production. If the Indians had directed the productive powers of the earth to the growth of corn, instead of to the growth of foxes' skins, they would have become rich. But they could not have reached this point without appropriation of the soil. They had learnt the necessity of appropriating the products of the soil, when they had bestowed labour upon obtaining them; but the last step towards productiveness was not taken. The Indians therefore were poor; the European settlers who had taken this last step were rich.

The want of resources in the country of the Indians, for the maintenance, even for a short time, of any considerable body of persons, is so notorious a fact, that when the Government of the United States, in 1802, gathered together the chiefs of the various tribes of the Creek Indians in their own country, to propose to them a plan for their civilization, it became necessary for the Commissioners of that Government to provide for the support of the people so assembled, by conveying food into the forests from the stores of the American towns.

The imperfect appropriation which exists amongst the Indians, preventing, as it does, the accumulation of capital, prevents the application of that skill and knowledge which is preserved and accumulated by the Division of employment. Tanner describes a poor fellow who was wounded in the arm by the accidental discharge of a gun. As there was little surgical skill

amongst the community, because no one could devote
himself to the business of surgery, the Indian, as the
only chance of saving his life, resolved to cut off his own
arm; "and taking two knives, the edge of one of which
he had hacked into a sort of saw, he with his right hand
and arm cut off his left, and threw it from him as far as
he could." The labour which an individual must go
through when the state of society is so rude that there
is scarcely any division of employment, and consequently
scarcely any exchanges, is exhibited in many passages of
Tanner's narrative. We select one. "I had no puk-
kavi, or mats for a lodge, and therefore had to build one
of poles and long grass. I dressed more skins, made my
own mocassins and leggings, and those for my children;
cut wood and cooked for myself and family, made my
snow-shoes, &c. &c. All the attention and labour I had
to bestow about home sometimes kept me from hunting,
and I was occasionally distressed for want of provisions.
I busied myself about my lodge in the night-time. When
it was sufficiently light I would bring wood, and attend
to other things without; at other times I was repairing
my snow-shoes, or my own or my children's clothes.
For nearly all the winter I slept but a very small part of
the night."

Tanner was thus obliged to do everything for himself,
and consequently to work at very great disadvantage,
because the principle of exchange was so imperfectly
acted upon by the people amongst whom he lived. This
principle of exchange was imperfectly acted upon, because
the principle of appropriation was imperfectly acted upon.
The occupation of all, and of each, was to hunt game, to
prepare skins, to sell them to the traders, to make sugar
from the juice of maple-trees, to build huts, and to sew
the skins which they dressed and the blankets which
they bought into rude coverings for their bodies. Every
one of them did all of these things for himself, and of
course he did them very imperfectly. The people were
not divided into hunters, and furriers, and dealers, and
sugar-makers, and builders, and tailors. Every man was
his own hunter, furrier, dealer, sugar-maker, builder,

and tailor; and consequently, every man, like Tanner, was so occupied by many things, that want of food and want of rest were ordinary sufferings. He describes a man who was so borne down and oppressed by these manifold wants, that in utter despair of being able to surmount them, he would lie still till he was at the point of starvation, replying to those who tried to rouse him to kill game, that he was too poor and sick to set about it. By describing himself as poor, he meant to say that he was destitute of all the necessaries and comforts whose possession would encourage him to add to the store. He had little capital. The skill which he possessed of hunting game gave him a certain command over the spontaneous productions of the forest; but, as his power of hunting depended upon chance supplies of game, his labour necessarily took so irregular a direction, and was therefore so unproductive, that he never accumulated sufficient for his support in times of sickness, or for his comfortable support at any time. He became, therefore, despairing; and had that perfect apathy, that indifference to the future, which is the most pitiable evidence of extreme wretchedness. This man felt his powerless situation more keenly than his companions; but with all savage tribes there is a want of steady and persevering exertion, proceeding from the same cause. Severe labour is succeeded by long fits of idleness, because their labour takes a chance direction. This is a universal case. Habits of idleness, of irregularity, of ferocity, are the characteristics of all those who maintain existence by the pursuit of the unappropriated productions of nature; while constant application, orderly arrangement of time, and civility to others, result from systematic industry. The savage and the poacher are equally the slaves of violent impulses—equally disgusted at the prospect of patient application. When the support of life depends upon chance supplies, the reckless spirit of a gambler is sure to take possession of the whole man; and the misery which results from these chance supplies produces either dejection or ferocity. The author of this book used to observe the habits of a class of such persons, who fre-

quent the Thames at Eton ; and he thus described them
in verses of his boyhood :—

> What boat is this which creeps so lazily
> Up the still stream ?   How quietly falls the drip
> Of the slow paddle!   Now it shoots along,
> As if that lone man fear'd us.   Well I ken
> His rough and dangerous trade.   He knows each hole
> Where the quick-sighted eel delights to swim
> When clouds obscure the moon ; and there he lays
> His traps and gins ; and then he sleeps awhile ;
> But rouses up before the prying dawn
> Betrays his course ; and out he cautiously glides
> To try his doubtful luck.   Perchance he finds
> Stores that may buy him bread ; but oft'ner still
> His toil is fruitless, and deject he comes
> Home to his emberless hearth, and sits him down,
> Idle and starving through the busy day.

Mungo Park describes the wretched condition of the
inhabitants of countries in Africa, where small particles
of gold are found in the rivers.   Their lives were spent
in hunting for the gold to exchange for useful commodities, instead of raising the commodities themselves ;—
and they were consequently poor and miserable, listless
and unsteady.   Their fitful industry had too much of
chance mixed up with it to afford a certain and general
profit.   The natives of Cape de la Hogue, in Normandy,
were the most wretched and ferocious people in all
France, because they depended principally for support
on the wrecks that were frequent on their coasts.   When
there were no tempests, they made an easy transition
from the character of wreckers to that of robbers.   A
benefactor of his species taught these unhappy people to
collect a marine plant to make potash.   They immediately became profitable labourers and exchangers ; they
obtained a property in the general intelligence of civilized
life ; the capital of society raised them from misery to
wealth, from being destroyers to being producers.

The Indians, as we thus see, are poor and wretched,
because they have no appropriation beyond articles of
domestic use ; because they have no property in land,

and consequently no cultivation. Yet even they are not insensible to the importance of the principle, for the preservation of the few advantages that belong to their course of life. Tanner says, " I have often known a hunter leave his traps for many days in the woods, without visiting them, or feeling any anxiety about their safety." The Indians even carry the principle of appropriation almost to a division of land ; for each tribe, and sometimes each individual, has an allotted hunting-ground— imperfectly appropriated, indeed, by the first comer, and often contested with violence by other hunters, but still showing that they approached the limit which divides the savage from the civilized state, and that if cultivation were introduced amongst them, there would be a division of land, as a matter of necessity. The security of individual property is the foundation of all social improvement. It is impossible to speak of the productive power of labour in the civilized state, without viewing it in connexion with that great principle of society which considers all capital as appropriated. If the Indians are ever civilized, their civilization will be the growth of their intercourse with the settlers of America and the traders of Europe. Their exchanges, in that case, will become more perfect. At present, the rude industry of the Indians is stimulated by the luxury of Europe into an employ which would cease to exist if the people became civilized. If agriculture were introduced amongst them—if they were to grow corn and keep domestic animals,—they would cease to be hunters of foxes and martens, because their wants would be much better supplied by other modes of labour, involving less suffering and less uncertainty. As it is, the traders, who want skins, do not think of giving the Indians tools to work the ground, and seeds to put in it, and cows and sheep to breed other cows and sheep. They avail themselves of the uncivilized state of these poor tribes, to render them the principal agents in the manufacture of fur, to supply the luxuries of another hemisphere. But still the exchange, imperfect as it is in all cases, and unjust as it is in many, is better for the Indians than no exchange ;

althougn we fear that ardent spirits take away from the Indians the greater number of the advantages which would otherwise remain with them as exchangers. If the Indians had no skins to give to Europe, Europe would have no blankets and ammunition to give to them. They would obtain their food and clothing by the use of the bow alone. They would live entirely from hand to mouth. They would have no motive for accumulation, because there would be no exchanges ; and they would consequently be even poorer and more helpless than they are now as exchangers of skins. They are suffering from the effects of small accumulations and imperfect exchange ; but these are far better than no accumulation and no exchange. If the course of their industry were to be changed by perfect appropriation—if they were consequently to become cultivators and manufacturers, instead of wanderers in the woods to hunt for wild and noxious animals—they would, in the course of years, have abundance of profitable labour, because they would have abundance of capital. As it is, their accumulations are so small, that they cannot proceed with their own uncertain labour of hunting without an advance of capital on the part of the traders ; and thus, even in the rude tradings of these poor Indians, credit, that complicated instrument of commercial exchange, operates upon the direction of their labour. Of course credit would not exist at all without appropriation. Their rights of property are perfect as far as they go ; but they are not carried far enough to direct their labour into channels which would ensure sufficient production for the labourers. Their labour is unproductive because they have small accumulations ;—their accumulations are small because they have imperfect exchange ;—their exchange is imperfect because they have limited appropriation. We may illustrate this state of imperfect production by another passage from Tanner's story :—

"The Hudson's Bay Company had now no post in that part of the country, and the Indians were soon made conscious of the advantage which had formerly resulted to them from the competition between rival trading companies. Mr. Wells, at the commencement of winter, called us all together, gave the Indians a ten-gallon keg of rum and some tobacco, telling them at the same time he would not credit one of them the value of a single needle. When they brought skins he would buy them, and give in exchange such articles as were necessary for their comfort and subsistence during the winter. I was not with the Indians when this talk was held. When it was reported to me, and a share of the presents offered me, I not only refused to accept anything, but reproached the Indians for their pusillanimity in submitting to such terms. They had been accustomed for many years to receive credits in the fall; they were now entirely destitute not of clothing merely, but of ammunition, and many of them of guns and traps. How were they, without the accustomed aid from the traders, to subsist themselves and their families during the ensuing winter? A few days afterwards, I went to Mr. Wells, and told him that I was poor, with a large family to support by my own exertions; and that I must unavoidably suffer, and perhaps perish, unless he would give me such a credit as I had always in the fall been accustomed to receive. He would not listen to my representation, and told me roughly to be gone from his house. I then took eight silver beavers, such as are worn by the women as ornaments on their dress, and which I had purchased the year before at just twice the price that was commonly given for a capote; * I laid them before him on the table, and asked him to give me a capote for them, or retain them as a pledge for the payment of the price of the garment, as soon as I could procure the peltries.† He took up the ornaments, threw them in my face, and told me never to come inside of his house again. The cold weather of the winter had not yet set in, and I went immediately to my hunting-ground, killed a number of moose, and set my wife to make the skins into such garments as were best adapted to the winter season, and which I now saw we should be compelled to substitute

* A sort of great-coat. † Skins.

c

for the blankets and woollen clothes we had been accustomed to receive from the traders."

This incident at once shows us that the great blessing
of the civilized state is its increase of the powers of production. Here we see the Indians, surrounded on all
sides by wild animals whose skins might be made into
garments, reduced to the extremity of distress because
the traders refused to advance them blankets and other
necessaries, to be used during the months when they
were employed in catching the animals from which they
might obtain the skins. It is easy to see that the Indians
were a long way removed from the power of making
blankets themselves. Before they could reach this point
their forests must have been converted into pasture-
grounds ;—they must have raised flocks of sheep, and
learnt all the various complicated arts, and possessed all
the ingenious machinery, for converting wool into cloth.
By their exchange of furs for blankets, they obtained a
share in the productiveness of civilization ;—they obtained comfortable clothing with much less labour than
they could have made it out of the furs. If Tanner had
not considered the capote which he desired to obtain
from the traders, better, and less costly, than the garment of moose-skins, he would not have carried on any
exchange of the two articles with the traders. The skins
of martens and foxes were only valuable to the Indians,
without exchange, for the purpose of sewing together to
make covering. They had a different value in Europe
as articles of luxury ; and therefore the Indians by exchange obtain a greater plenty of superior clothing than
if they had used the skins themselves. But the very
nature of the trade, depending upon chance supplies,
rendered it impossible that they should accumulate.
They had such pressing need of ammunition, traps, and
blankets, that the produce of the labour of one hunting
season was not more than sufficient to procure the commodities which they required to consume in the same
season. But supposing the Indians could have bred
foxes and martens and beavers, as we breed rabbits, for
the supply of the European demand for fur, doubtless

they would have then advanced many steps in the character of producers. The thing is perhaps impossible; but were it possible, and were the Indians to have practised it, they would immediately have become capitalists, to an extent that would have soon rendered them independent of the credit of the traders. They must, however, have previously established a more perfect appropriation. Each must have enclosed his own hunting-ground ; and each must have raised some food for the maintenance of his own stock of beavers, foxes, and martens. It would be easier, doubtless, to raise the food for themselves, and ultimately to exchange corn for clothing, instead of furs for clothing. When this happens—and it will happen sooner or later, unless the Indians are extirpated by their poverty which proceeds from their imperfect production—Europe must go without the brilliant variety of skins which we procure at the cost of so much labour, accompanied with so much wretchedness, because the labour is so unproductive to the labourers. When the ladies of London and Paris are compelled to wear tippets of rabbits instead of sables, and when the hair of the beaver ceases to be employed in the manufacture of our hats, the woods of America will have been cut down —corn will be growing in their place—the inhabitants will be weaving cloth instead of trapping foxes ;—there will be appropriation and capital, profitable labour and comfort.

CHAPTER IV.

THERE is an account in Foster's Essays (a work of great merit) of a man who, having by a short career of boundless extravagance dissipated every shilling of a large estate which he inherited from his fathers, obtained possession again of the whole property by a course which the writer well describes as a singular illustration of decision of character. The unfortunate prodigal, driven forth from the home of his early years by his own imprudence, and reduced to absolute want, wandered about for some time in a state of almost unconscious despair, meditating self-destruction, till he at last sat down upon a hill which overlooked the fertile fields that he once called his own. " He remained," says the narrative, " fixed in thought a number of hours, at the end of which he sprang from the ground with a vehement exulting emotion. He had formed his resolution, which was, that all these estates should be his again; he had formed his plan, too, which he instantly began to execute." We shall show, by and by, how this plan worked in detail;—it will be sufficient, just now, to examine the principles upon which it was founded. He looked to no freak of fortune to throw into his lap by chance what he had cast from him by wilfulness. He neither trusted to inherit those lands from their present possessor by his favour, nor to wring them from him by a course of law. He was not rash and foolish enough to dream of obtaining again by force those possessions which he had exchanged for vain superfluities. But he resolved to become once more their master by the operation of the only principle which could give them to him in a civilized society. He resolved to obtain them again by the same agency through which he had lost them—by exchange. But what had he to exchange? His capital was gone, even to the uttermost farthing; he must labour to obtain new capital. With a courage worthy of imitation he resolved to accept the very first work that should be offered to him, and, however low the wages of that work, to spend only a part of those wages, leaving something for a store. The day that he made this resolution he carried it into execution. He found some service to be performed—irksome, doubtless, and in many eyes degrading. But he had a purpose which made every occupation appear honourable, as every occupation truly is that is productive of utility. Incessant labour and scrupulous parsimony soon accumulated for him a capital; and the store, gathered together with such energy, was a rapidly increasing one. In no very great number of years the once destitute labourer was again a rich proprietor. He had earned again all that he had lost. The lands of his fathers were again his. He had accomplished his plan.

A man so circumstanced—one who possesses no capital, and is only master of his own natural powers—if suddenly thrown down from a condition of ease, must look upon the world, at the first view, with deep apprehension. He sees everything around him appropriated. He is in the very opposite condition of Alexander Selkirk, when he is made to exclaim " I am monarch of all I survey." Instead of feeling that his " right there is none to dispute," he knows that every blade of corn that covers the fields, every animal that grazes in the pastures, is equally numbered as the property of some individual owner, and can only pass into his possession by exchange. In the towns it is the same as in the country. The dealer in bread and in clothes,—the victualler from whom he would ask a cup of beer and a night's lodging,—will give him nothing, although they will exchange everything. He cannot exist, except as a beggar, unless he puts himself in the condition to become an exchanger.

But still, with all these apparent difficulties, his prospects of subsisting, and of subsisting comfortably, are far greater than in any other situation in which he must labour to live. As we have already seen, the condition of by far the greater number of the millions that constitute the exchangers of civilized society is greatly superior to that of the few thousands who exist upon the precarious supplies of the unappropriated productions of nature in the savage life. Although an exchange must always be made—although in very few cases "the fowl and the brute" offer themselves to the wayfaring man for his daily food—although no herbs worth the gathering can be found for the support of life in the few uncultivated parts of a highly cultivated country—the aggregate riches are so abundant, and the facilities which exist for exchanging capital for labour are therefore so manifold, that the poorest man in a state of civilization has a much greater certainty of supplying all his wants, and of supplying them with considerably more ease, than the richest man in a state of uncivilization. The principle upon which he has to rely is, that in a highly civilized country there is large production. There is large production because there is profitable labour;—there is profitable labour because there is large accumulation;—there is large accumulation because there is unlimited exchange; —there is unlimited exchange because there is universal appropriation. John Tanner was accounted a rich man by the Indians—doubtless because he was more industrious than the greater number of them; but we have seen what privations he often suffered. He suffered privations because there was no capital, no accumulation of the products of labour, in the country in which he lived. Where such a store exists, the poorest man has a tolerable certainty that he may obtain his share of it as an exchanger; and the greater the store the greater the certainty that his labour, or power of adding to the store, will obtain a full proportion of what previous labour has gathered together.

It may assist us in making the value of capital more clear, if we take a rapid view of the most obvious features of the accumulation of a highly civilized country.

The first operation in a newly settled country is what is termed to clear it. Look at a civilized country, such as England. It *is* cleared. The encumbering woods are cut down, the unhealthy marshes are drained. The noxious animals which were once the principal inhabitants of the land are exterminated; and their place is supplied with useful creatures, bred, nourished, and domesticated by human art, and multiplied to an extent exactly proportioned to the wants of the population. Forests remain for the produce of timber, but they are confined within the limits of their utility;—mountains " where the nibbling flocks do stray," have ceased to be barriers between nations and districts. Every vegetable that the diligence of man has been able to transplant from the most distant regions is raised for food. The fields are producing a provision for the coming year; while the stock for immediate consumption is ample, and the laws of demand and supply are so perfectly in action, that scarcity seldom occurs, and famine never. Rivers have been narrowed to bounds which limit their inundations, and they have been made navigable wherever their navigation could be profitable. The country is covered with roads, with canals, and now, more especially, with railroads, which render distant provinces as near to each other for commercial purposes as neighbouring villages in less advanced countries. Houses, all possessing some comforts which were unknown even to the rich a few centuries ago, cover the land, in scattered farm-houses and mansions, in villages, in towns, in cities, in capitals. These houses are filled with an almost inconceivable number of conveniences and luxuries—furniture, glass, porcelain, plate, linen, clothes, books, pictures. In the stores of the merchants and traders, the resources of human ingenuity are displayed in every variety of substances and forms that can exhibit the multitude of civilized wants; and in the manufactories are seen the wonderful adaptations of science for satisfying those wants at the cheapest cost. The people who inhabit such a civilized land have not only the readiest communication with each other by the means of roads and

canals, but can trade by the agency of ships with all
parts of the world. To carry on their intercourse amongst
themselves they speak one common language, reduced to
certain rules, and not broken into an embarrassing variety
of unintelligible dialects. Their written communications
are conveyed to the remotest corners of their own
country, and even to other kingdoms, with the most
unfailing regularity, and now with a cheapness which
makes the poorest and the richest equal in their power
to connect the distant with their thoughts by written communication. Whatever is transacted in such a populous
hive, the knowledge of which can afford profit or amusement to the community, is recorded with a rapidity which
is not more astonishing than the general accuracy of the
record. What is more important, the discoveries of
science, the elegancies of literature, and all that can
advance the general intelligence, are preserved and
diffused with the utmost ease, expedition, and security,
so that the public stock of knowledge is constantly increasing. Lastly, the general well-being of all is sustained
by laws,—sometimes indeed imperfectly devised and expensively administered, but on the whole of infinite value
to every member of the community ; and the property of
all is defended from external invasion and from internal
anarchy by the power of government, which will be
respected only in proportion as it advances the general
good of the humblest of its subjects, by securing their
capital from plunder and defending their industry from
oppression.

This capital is ready to be won by the power of the
working man. But he must exercise this power in complete subjection to the natural laws by which every
exchange of society is regulated. These laws sometimes
prevent labour being instantly exchanged with capital,
for an exchange necessarily requires a balance to be preserved between what one man has to supply and what
another man has to demand ; but in their general effect
they secure to labour the certainty that there shall be
abundance of capital to exchange with ; and that if
prudence and diligence go together, the labourer may

himself become a capitalist, and even pass out of the condition of a labourer into that of a proprietor, or one who lives upon accumulated produce. The experience of every day shows this process going forward—not in a solitary instance, as in that of the ruined and restored man whose tale we have just told, but in the case of thriving tradesmen all around us, who were once work-men. But if the labourer or the great body of labourers were to imagine that they could obtain such a proportion of the capital of a civilized country except as exchangers, the store would instantly vanish. They might perhaps divide by force the crops in the barns and the clothes in warehouses—but there would be no more crops or clothes. The principle upon which all accumulation depends, that of security of property, being destroyed, the accumulation would be destroyed.

When we look at the nature of the accumulated wealth of society, it is easy to see that the poorest member of it who dedicates himself to profitable labour is in a certain sense rich—rich, as compared with the unproductive and therefore poor individuals of any uncivilized tribe. The very scaffolding, if we may so express it, of the social structure, and the moral forces by which that structure was reared, and is upheld, are to him riches. To be rich is to possess the means of supplying our wants—to be poor is to be destitute of those means. Riches do not consist only of money and lands, of stores of food or clothing, of machines and tools. The particular know-ledge of any art,—the general understanding of the laws of nature,—the habit from experience of doing any work in the readiest way,—the facility of communicating ideas by written language,—the enjoyment of institutions con-ceived in the spirit of social improvement,—the use of the general conveniences of civilized life, such as roads—these advantages, which the poorest man in England possesses or may possess, constitute individual property. They are means for the supply of wants, which in them-selves are essentially more valuable for obtaining his full share of what is appropriated, than if all the productive powers of nature were unappropriated, and if, consequently, these great elements of civilization did not exist. Society obtains its almost unlimited command over riches by the increase and preservation of knowledge, and by the division of employments, including union of power. In his double capacity of a consumer and a producer, the humblest man has the full benefit of these means of wealth—of these great instruments by which the productive power of labour is carried to its highest point.

But if these common advantages, these public means of society, offering so many important agents to the individual for the gratification of his wants, alone are worth more to him than all the precarious power of the savage state,—how incomparably greater are his advantages when we consider the wonderful accumulations, in the form of private wealth, which are ready to be exchanged with the labour of all those who are in a condition to add to the store. It has been truly said, " It is a great misfortune to be poor, but it is a much greater misfortune for the poor man to be surrounded only with other poor like himself."* The reason is obvious. The productive power of labour can be carried but a very little way without accumulation of capital. In a highly civilized country, capital is heaped up on every side by ages of toil and perseverance. A succession, during a long series of years, of small advantages to individuals unceasingly renewed and carried forward by the principle of exchanges, has produced this prodigious amount of the aggregate capital of a country whose civilization is of ancient date. This accumulation of the means of existence, and of all that makes existence comfortable, is principally resulting from the labours of those who have gone before us. It is a stock which was beyond their own immediate wants, and which was not extinguished with their lives. It is our capital. It has been produced by labour alone, physical and mental. It can be kept up only by the same power which has created it, carried to the highest point of productiveness by the arrangements of society.

* Say.

CHAPTER V.

THERE is an old proverb, that "When two men ride on one horse, one man must ride behind." Capital and Labour are, as we think, destined to perform a journey together to the end of time. We have shown how they proceed on this journey. We have shown that although Labour is the parent of all wealth, its struggles for the conversion of the rude supplies of nature into objects of utility, are most feeble in their effects till they are assisted by accumulation. Before the joint interests of Labour and Capital were at all understood, they kept separate; when they only began to be understood, as we shall show, they were constantly pulling different ways, instead of giving "a long pull, a strong pull, and a pull altogether;" and even now, when these interests in many respects are still imperfectly understood, they occasionally quarrel about the conditions upon which they will continue to travel in company. In the very outset of the journey, Labour, doubtless, took the lead. In the dim morning of society Labour was up and stirring before Capital was awake. Labour did not then ride; he travelled slowly on foot through very dirty ways. Capital, at length, as slowly followed after, through the same mire, but at an humble distance from his parent. But when Capital grew into strength, he saw that there were quicker and more agreeable modes of travelling for both, than Labour had found out. He procured that fleet and untiring horse Exchange; and when he proposed to Labour that they should mount together, he claimed the right, and kept it, for their mutual benefit, of taking the direction of the horse. For this reason, as it appears to us, we are called upon to assign to one of the companions, according to the practice of the old Knights Templars, the privilege of sitting before the other—holding the reins, indeed, but in all respects having a community of interests, and an equality of duties, as well as rights, with his fellow-traveller.

Let us endeavour to advance another step in the illustration of these positions, by going back to the prodigal who had spent all his substance. Let us survey him at the moment when he had made the wise, and in many respects heroic, resolution to pass from the condition of a consumer into that of a producer. The story says, " The first thing that drew his attention was a heap of coals shot out of a cart on the pavement before a house. He offered himself to shovel or wheel them into the place where they were to be laid, and was employed." Here, then, we see that the labour of this man was wholly and imperatively directed by accumulation. It was directed as absolutely by the accumulation of others as the labour of Dampier's Moskito Indian was directed by his own accumulation. The Indian could not labour profitably—he could not obtain fish and goats for his food, instead of seal's flesh—till he had called into action the power which he possessed in his knife and his gun-barrel. The prodigal had no accumulation whatever of his own. He had not even the accumulation of peculiar skill in any mode of labour ;—for a continual process of waste enlarges neither the mental nor physical faculties, and leaves the wretched being who has to pass into the new condition of a producer as helpless as the weakest child. He had nothing but the lowest power, of labouring without peculiar knowledge or skill. He had, however, an intensity and consistency of purpose which raised this humble power into real strength. He was determined never to go backward—always to go on. He knew, too, his duties as well as his rights ; and he saw that he must wholly accommodate his power to the greater power which was in action around him. When he passed into the condition of a producer, he saw that his powers and rights were wholly limited and directed by the principles necessary to advance production; and that his own share of what he assisted in producing must be measured by the laws which enabled him to produce at all. He found himself in a position where his labour was absolutely governed by the system of exchanges. No other system could operate around him; because he was in a civilized country. Had he been thrown upon a desert land without food and shelter, his labour must have been instantly and directly applied to procuring food and shelter. He was equally without food and shelter in a civilized country. But the system of exchanges being in action, he did not apply his labour directly to the production of food and lodging for himself. He added by his labour a new value to a heap of coals; he enabled another man more readily to acquire the means of warmth; and by this service, which he exchanged for " a few pence" and " a small gratuity of meat and drink," he indirectly obtained food and lodging. He conferred an additional value upon a heap of coals; and that additional value was represented by the " few pence" and " a small gratuity of meat and drink." Had the system of exchange been less advanced, that is, had society been less civilized, he would probably have exchanged his labour for some object of utility, by another and a ruder mode. He would have received a portion of the coals as the price of the labour by which he gave an additional value to the whole heap. But mark the inconvenience of such a mode of exchange. His first want was food; his next, shelter: had he earned the coals, he must have carried them about till he had found some other person ready to exchange food and lodging for coals. Such an occurrence might have happened, but it would have been a lucky accident. He could find all persons ready to exchange food and clothes for money —because money was ready again to exchange for other articles of utility which they might require, and which they would more readily obtain by the money than by the food and clothes which our labourer had received for them. During the course of the unprofitable labour of waiting till he had found an exchanger who wanted

coals, he might have perished. What then gave him
the means of profitable labour, and furnished him with
an article which every one was ready to receive in exchange for articles of immediate necessity? Capital in
two forms. The heap of coals was capital. The coals
represented a very great and various accumulation of
former labour that had been employed in giving them
value. The coals were altogether valueless till labour
had been employed to raise them from the pit, and to
convey them to the door of the man who was about to
consume them. But with what various helps had this
labour worked! Mere manual labour could have done
little or nothing with the coals in the pit. Machines
had raised them from the pit. Machines had transported
them from the pit to the door of the consumer. They
would have remained buried in the earth but for large
accumulations of knowledge, and large accumulations of
pecuniary wealth to set that knowledge in action by exchanging with it. The heap of coals represented all this
accumulation; and it more immediately represented the
Circulating Capital of consumable articles of utility,
which had been paid in the shape of wages, at every
stage of the labour exercised in raising the coals from
the mine, and conveying them to the spot in which the
prodigal found them laid. The coals had almost attained
their highest value by a succession of labour; but one
labour was still wanting to give them the highest value.
They were at their lowest value when they remained
unbroken in the coal-pit; they were at their highest
value when they were deposited in the cellar of the consumer. For that last labour there was circulating capital
ready to be exchanged. The man whose course of production we have been tracing imparted to them this last
value, and for this labour he received a "few pence"
and a "gratuity of meat and drink." These consumable
commodities, and the money which might be exchanged
for other consumable commodities, were circulating capital. They supplied his most pressing wants with
incomparably more readiness and certainty than if he
had been turned loose amongst the unappropriated pro-

ductions of nature, with unlimited freedom and absolute rights. In the state in which he was actually placed his rights were limited by his duties,—but this balance of rights and duties was the chief instrument in the satisfaction of his wants. Let us examine the principle a little more in detail.

An exchange was to be carried on between the owner of the coals and the man who was willing to shovel them into the owner's cellar. The labourer did not want any distinct portion of the coals, but he wanted some articles of more urgent necessity in exchange for the new value which he was ready to bestow upon the coals. The object of each exchanger was, that labour should be exchanged with capital. That object could not have been accomplished, or it would have been accomplished slowly, imperfectly, and therefore unprofitably, unless there had been interchangeable freedom and security for both exchangers,—for the exchanger of capital, and the exchanger of labour. The first right of the labourer was, that his labour should be free ;—the first right of the capitalist was, that his capital should be free. The rights of each were built upon the security of property. Could this security have been violated, it might have happened, either that the labourer should have been compelled to shovel in the coals—or, that the capitalist should have been compelled to employ the labourer to shovel them in. Had the lot of the unfortunate prodigal been cast in such a state of society as would have allowed this violation of the natural rights of the labourer and the capitalist, he would have found little accumulation to give a profitable direction to his labour. He would have found production suspended, or languishing. There would probably have been no heap of coals wanting his labour to give them the last value ;—for the engines would have been idle that raised them from the pit, and the men would have been idle that directed the engines. The circulating capital that found wages for the men, and fuel for the engines, would have been idle, because it could not have worked with security. Accumulation, therefore, would have been suspended ;—and all profitable labour would, in consequence, have been suspended. It was the unquestionable right of the labourer that his labour should be free; but it was balanced by the right of the capitalist that his accumulation should be secure. Could the labour have seized upon the capital, or the capital upon the labour, production would have been stopped altogether, or in part. The mutual freedom and security of labour and capital compel production to go forward; and labour and capital take their respective stations, and perform their respective duties, altogether with reference to the laws which govern production. These laws are founded upon the natural action of the system of exchange, carrying forward all its operations by the natural action of the great principle of demand and supply. When capital and labour know how to accommodate themselves to the direction of these natural laws, they are in a healthy state with respect to their individual rights, and the rights of industry generally. They are in that state in which each is working to the greatest profit in carrying forward the business of production.

The story of the prodigal goes on to say, " He then looked out for the next thing that might chance to offer; and went with indefatigable industry through a succession of servile employments, in different places, of longer and shorter duration." By following the direction which capital gave to his industry, he became at length a capitalist. " He had gained, after a considerable time, money enough to purchase, in order to sell again, a few cattle, of which he had taken pains to understand the value. He speedily, but cautiously, turned his first gains into second advantages; retained, without a single deviation, his extreme parsimony; and thus advanced by degrees into larger transactions and incipient wealth. The final result was, that he more than recovered his lost possessions, and died an inveterate miser, worth 60,000l."

CHAPTER VI.

If we have succeeded in making our meaning clear, by stating a general truth, not in an abstract form, but as brought out by various instances of the modes in which it is exhibited, we shall have led the reader to the conclusion that accumulation, or capital, is absolutely essential to the profitable employment of labour ; and that the greater the accumulation, the greater the extent of that profitable employment. This truth, however, has been denied altogether by some speculative writers ;—and, what is more important, has been practically denied by the conduct of nations and individuals in the earlier state of society,—and is still denied by existing prejudices, derived from the current maxims of former days of ignorance and half-knowledge. With the speculative writers we have little to do. When Rousseau, for instance, advises governments not to secure property to its possessors, but to deprive them of all means of accumulating, it is sufficient to know that the same writer advocated the savage state, in which there should be no property, in preference to the social, which is founded on appropriation. Knowing this, and being convinced that the savage state, even with imperfect appropriation, is one of extreme wretchedness, we may safely leave such opinions to work their own cure. For it is not likely that any individual, however disposed to think that accumulation is an evil, would desire, by destroying accumulation, to pass into the condition, described by John Tanner, of a constant encounter with hunger in its most terrific forms : and seeing, therefore, the fallacy of such an opinion, he will also see that if he partially destroys accumulation, he equally impedes production, and equally destroys his share in the productive power of capital and labour working together for a common good in the social state.

But, without going the length of wishing to destroy capital, there are many who think that accumulation is a positive evil, and that consumption is a positive benefit; and, therefore, that economy is an evil, and waste a benefit. The course of a prodigal man is by many still viewed with considerable admiration. He sits up all night in frantic riot—he consumes whatever can stimulate his satiated appetite—he is waited upon by a crowd of unproductive and equally riotous retainers—he breaks and destroys everything around him with an unsparing hand—he rides his horses to death in the most extravagant attempts to wrestle with time and space; and when he has spent all his substance in these excesses, and dies an outcast and a beggar, he is said to have been a hearty fellow, and to have " made good for trade." When, on the contrary, a man of fortune economizes his revenue— lives like a virtuous and reasonable being, whose first duty is the cultivation of his understanding—eats and drinks with a regard to his health—keeps no more retainers than are sufficient for his proper comfort and decency—breaks and destroys nothing—has respect to the inferior animals, as well from motives of prudence as of mercy—and dies without a mortgage on his lands; *he* is said to have been a stingy fellow, who did not know how to " circulate his money." To " circulate money," to " make good for trade," in the once common meaning of the terms, is for *one* to consume unprofitably what, if economized, would have stimulated production in a way that would have enabled *hundreds*, instead of one, to consume profitably. We will give you two historical examples of these two opposite modes of making good for trade, and circulating money. The Duke of Buckingham, " having been possessed of about 50,000*l.* a year, died in 1687, in a remote inn in Yorkshire, reduced to the utmost misery."* After a life of the most

* Ruffhead's Pope.

wanton riot, which exhausted all his princely resources, he was left at the last hour, under circumstances which are well described in the following lines by Pope :—

" In the worst inn's worst room, with mat half hung,
  The floors of plaster, and the walls of dung,
  On once a flock bed, but repaired with straw,
  With tape-tied curtains never meant to draw,
  The George and Garter dangling from that bed
  Where tawdry yellow strove with dirty red;
  Great Villiers lies.   .   .   .   ,
  No wit to flatter left of all his store,
  No fool to laugh at, which he valued more,
  There, victor of his health, of fortune, friends
  And fame, this lord of useless thousands ends."

Contrast the course of this unhappy man, with that of the Duke of Bridgewater, who devoted his property to really " making good for trade," by constructing the great canals which connect Manchester with the coal countries and with Liverpool. The Duke of Buckingham lived in a round of sensual folly : the Duke of Bridgewater limited his personal expenditure to 400*l.* a year, and devoted all the remaining portion of his revenues to the construction of a magnificent work of the highest public utility.   The one supported a train of cooks and valets and horse-jockeys : the other called into action the labour of thousands, and employed in the direction of that labour the skill of Brindley, the greatest engineer that any country has produced. The one died without a penny, loaded with debt, leaving no trace behind him but the ruin which his waste had produced : the other bequeathed almost the largest property in Europe to his descendants, and opened a channel for industry which afforded, and still affords, employment to thousands.

When we hear of a mob breaking windows on an illumination night, we are too apt to say they have " made good for trade." Is it not evident that the capital which was represented by the unbroken windows is really so much destroyed of the national riches, when the windows are broken ?—for if the windows had re-

mained unbroken, the capital would have remained to
stimulate the production of some new object of utility.
The glaziers, indeed, replaced the windows; but there
having been a destruction of the windows, there must
have been a necessary retrenchment in some other outlay, that would have afforded benefit to the consumer.
Doubtless, when the glazier is called into activity by a
mob breaking windows, some other trade suffers; for
the man who has to pay for the broken windows must
retrench somewhere, and, if he has less to lay out, some
other person has less to lay out. The glass-maker, probably, makes more glass at the moment; but he does so
to exchange with the capital that would otherwise have
gone to the maker of clothes or of furniture : and, there
being an absolute destruction of the funds for the maintenance of labour, by an unnecessary destruction of what
former labour has produced, trade generally is injured to
the extent of the destruction. Some say that a fire
makes good for trade. The only difference of evil between the fire which destroys a house, and the mob
which breaks the windows, is, that the fire absorbs capital for the maintenance of trade, or labour, in the proportion of a hundred to one, when compared with the
mob. Some say that war makes good for trade. The
only difference of pecuniary evil (the moral evils admit
of no comparison) between the fire and the war is, that
the war absorbs capital for the maintenance of trade or
labour, in the proportion of a million to a hundred,
when compared with the fire. If the incessant energy
of production were constantly repressed by mobs, and
fires, and wars, the end would be that consumption
would altogether exceed production; and that then the
producers and the consumers would both be starved into
wiser courses, and perceive that nothing makes good for
trade but profitable industry and judicious expenditure.
Prodigality devotes itself too much to the satisfaction of
present wants ; avarice postpones too much the present
wants to the possible wants of the future. Real economy
is the happy measure between the two extremes ; and
that only "makes good for trade," because, while it car-

ries on a steady demand for industry, it accumulates a portion of the production of a country to stimulate new production. That judicious expenditure consists in

" The sense to value riches, with the art
T' enjoy them."

The fashion of "making good for trade" by unprofitable consumption is a relic of the barbarous ages. In the twelfth century, a count of France commanded his vassals to plough up the soil round his castle, and he sowed the ground with coins of gold, to the amount of fifteen hundred guineas, that he might have all men talk of his magnificence. Piqued at the lordly prodigality of his neighbour, another noble ordered thirty of his most valuable horses to be tied to a stake and burnt alive, that he might exhibit a more striking instance of contempt for accumulation. In the latter part of the fourteenth century, a Scotch noble, Colin Campbell, on receiving a visit from the O'Neiles of Ireland, ostentatiously burnt down his house at Inverary upon their departure ; and an Earl of Athol pursued the same course in 1528, after having entertained the papal Legate, upon the pretence that it was "the constant habitude of the highlanders to set on fire in the morning the place which had lodged them the night before." When the feudal lords had so little respect for their own property, it was not likely that they would have much regard for the accumulation of others. The Jews, who were the great capitalists of the middle ages, and who really merit the gratitude of Europeans for their avarice, as that almost alone enabled any accumulation to go forward, and any production to increase, were, as it is well known, persecuted in every direction by the crown, by the nobles, by the people. When a solitary farmer or abbot attempted to accumulate corn, which accumulation could alone prevent the dreadful famines invariably resulting from having no stock that might be available upon a bad harvest, the people burnt the ricks of the provident men, by way of lessening the evils of scarcity. The consequence was, that no person thought of accumulating at all, and that the price of wheat often rose, just before the harvest, from five shillings a quarter to five pounds.

During these dark periods the crown carried on the war against capital with an industry that could not be exceeded by that of the nobles or the people. Before the great charter the Commons complained that King Henry seized upon whatever was suited to his royal pleasure—horses, implements, food, anything that presented itself in the shape of accumulated labour. In the reign of Henry III. a statute was passed to remedy excessive distresses; from which it appeared that it was no unfrequent practice for the king's officers to take the opportunity of seizing the farmer's oxen, at the moment when they were employed in ploughing, or, as the statute says, "winning the earth,"—taking them off, and starving them to death, or only restoring them with new and enormous exactions for their keep. Previous to the Charter of the Forest no man could dig a marl-pit on his own ground, lest the king's horses should fall into it when he was hunting. As late as the time of James I. we find, from a speech of the great Lord Bacon, that it was a pretty constant practice of the king's purveyors to extort large sums of money by threatening to cut down favourite trees which grew near a mansion-house or in avenues. Despotism, in all ages, has depopulated the finest countries, by rendering capital insecure, and therefore unproductive; insomuch that Montesquieu lays it down as a maxim, that lands are not cultivated in proportion to their fertility, but in proportion to their freedom. In the fifteenth century, in England, we find sums of money voted for the restoration of decayed towns and villages. Just laws would have restored them much more quickly and effectually. The state of agriculture was so low that the most absurd enactments were made to compel farmers to till and sow their own lands, and calling upon every man to plant at least forty beans. All the laws for the regulation of labourers, at the same period, assumed that they should be *compelled* to work, and not wander about the country,—just in the same way that farmers should be compelled to sow and till.

It is perfectly clear that the towns would not have been depopulated, and gone to decay, if the accumulation of capital had not been obstructed by insecurity and wasted by ignorance; and that the same insecurity, and the same waste, rendered it necessary to assume that the farmer would not till and sow, and the labourer would not labour, without compulsion. The natural stimulus to industry was wanting, and therefore there was no industry, or, only unprofitable industry. The towns decayed, the country was uncultivated—production languished—the people were all poor and wretched; and the government dreamt that acts of parliament and royal ordinances could rebuild the houses and cultivate the land, when the means of building and cultivation, namely, the capital of the country, was exhausted by injustice producing insecurity.

But if the king, the nobles, and the people of the middle ages conspired together, or acted at least as if they conspired, to prevent the accumulation of capital, the few capitalists themselves, by their monstrous regulations, which, unfortunately, still throw their dark shadows over our own days, prevented that freedom of industry without which capital could not accumulate. Undoubtedly, the commercial privileges of corporations originally offered some barriers against the injustice of the crown and of the nobility; but the good was always accompanied with an evil, which rendered it, to a certain extent, valueless. Where these privileges gave security, they were a good; where they destroyed freedom, they were an injury. Instead of encouraging the intercourse between one trade and another, they encircled every trade with the most absurd monopolies and exclusive privileges. Instead of rendering commerce free between one district and another, they, even in the same country, encompassed commerce with the most harassing restrictions, which separated county from county, and town from town, as if seas ran between them. If a man of Coventry came to London with his wares, he was encountered at every step with the privileges of companies; if the man of London sought to

trade at Coventry, he was obstructed by the same corporate rights, preventing either the Londoner or the
Coventry man trading with advantage. The revenues
of every city were derived from forfeitures upon trades,
almost all established upon the principle that, if one
trade became too industrious, or too clever, it would be
the ruin of another trade. Every trade was fenced
round with secrets; and the commonest trade, as we
know from the language of an apprentice's indenture,
was called an "art and mystery." All these follies
went upon the presumption that "one man's gain is another man's loss," instead of vanishing before the truth,
that, in proportion as the industry of all men is free, so
will it be productive; and that production on all sides
ensures a state of things in which every exchanger is a
gainer, and no one a loser.

It is not to be wondered at that, while such opinions
existed, the union of capital and labour should have been
very imperfect; and that, while the capitalists oppressed
the labourers, in the same way that they oppressed each
other, the labourers should have thought it not unreason-
able to plunder the capitalists. It is stated by Harrison,
an old writer of credit,* that, during the single reign of
Henry VIII., seventy-two thousand thieves were hanged
in England. No fact can exhibit in a stronger light the
universal misery that must have existed in those days.
The whole kingdom did not contain half a million
grown-up males, so that, considering that the reign of
Henry VIII. extended over two generations, about one
man in ten must have been, to use the words of the same
historian, "devoured and eaten up by the gallows." In
the same reign the first statute against Egyptians (gip-
sies) was passed. These people went from place to
place in great companies—spoke a cant language, which
Harrison calls Pedlar's French—and were subdivided
into fifty-two different classes of thieves. The same race
of people prevailed throughout Europe. Cervantes, the
author of 'Don Quixote,' says of the Egyptians, or

* Preface to the Chronicles of Holingshed, an historian of
the time of Elizabeth.

Bohemians, that they seem to have been born for no other purpose than that of pillaging. While this universal plunder went forward, it is evident that the insecurity of property must have been so great that there could have been little accumulation, and therefore little production. Capital was destroyed on every side; and, because profitable labour had become so scarce by the destruction of capital, one-half of the community sought to possess themselves of the few goods of the other half, not as exchangers, but as robbers. As the robbers diminished the capital, the diminution of capital increased the number of robbers; and if the unconquerable energy of human industry had not gone on producing, slowly and painfully indeed, but still producing, the country would have soon fallen back to the state in which it was a thousand years before, when wolves abounded more than men. One great cause of all this plunder and misery was the oppression of the labourers.

CHAPTER VII.

ADAM Smith, in his great work, 'The Wealth of Nations,' says, "The property which every man has in his own labour, as it is the original foundation of all other property, so it is the most sacred and inviolable. The patrimony of a poor man lies in the strength and dexterity of his hands; and to hinder him from employing this strength and dexterity in what manner he thinks proper, without injury to his neighbour, is a plain violation of this most sacred property." The right of property, in general, has been defined by another writer, M. Say, to be "the exclusive faculty guaranteed to a man, or body of men, to dispose, at their own pleasure, of that which belongs to them." There can be no doubt that labour is entitled to the same protection as a property that capital is entitled to. There can be no doubt that the labourer has rights over his labour which no government and no individual should presume to interfere with. There can be no doubt that, as an exchanger of labour for capital, the labourer ought to be assured that the exchange shall in all respects be as free as the exchanges of any other description of property. His rights as an exchanger are, that he shall not be compelled to part with his property, by any arbitrary enactments, without having as ample an equivalent as the general laws of exchange will afford him; that he shall be free to use every just means, either by himself or by union with others, to obtain such an equivalent; that he shall be at full liberty to offer that property in the best market that he can find, without being limited to any particular market; that he may give to that property every modification which it is capable of receiving from his own natural or acquired skill, without being narrowed to any one form of producing it. In other words, natural justice demands that the working-man shall work when he please, and be idle when he please, always providing that if he make a contract to work, he shall not violate that engagement by remaining idle ; that no labour shall be forced from him, and no rate of payment for that labour prescribed by statutes or ordinances ; that he shall be free to obtain as high wages as he can possibly get, and unite with others to obtain them, always providing that in his union he does not violate that freedom of industry in others which is the foundation of his own attempts to improve his condition ; that he may go from place to place to exchange his labour without being interfered with by corporate rights or monopolies of any sort, whether of masters or workmen ; and that he may turn from one employment to the other, if he so think fit, without being confined to the trade he originally learnt, or may strike into any line of employment without having regularly learnt it at all. When the working-man has these rights secured to him by the sanction of the laws, and the concurrence of the institutions and customs of the country in which he lives, he is in the condition of a free exchanger. He has the full, uninterrupted, absolute possession of his property. He is upon a perfect legal equality with the capitalist. He may labour cheerfully with the well-founded assurance that his labour will be profitably exchanged for the goods which he desires for the satisfaction of his wants, as far as laws and institutions can so provide. In a word, he may assure himself that if he possesses anything valuable to offer in exchange for capital, the capital will not be fenced round with any artificial barriers, or invested with any unnatural preponderance, to prevent the exchange being one of perfect equality, and therefore a real benefit to both exchangers.

We are approaching this desirable state in England. Indeed, there is scarcely any legal restriction acted upon which prevents the exchange of labour with capital being completely unembarrassed. Yet it is only within a few centuries that the working-men of this country have emerged from the condition of actual slaves into that of free labourers ; it is only a few hundred years ago since the cultivator of the ground, the domestic servant, and sometimes even the artisan, was the absolute property of another man—bought, sold, let, without any will of his own, like an ox or a horse—producing nothing for himself—and transmitting the miseries of his lot to his children. The progress of civilization destroyed this monstrous system, in the same way that at the present day it is destroying it in Russia and other countries where slavery still exists. But it was by a very slow process that the English slave went forward to the complete enjoyment of the legal rights of a free exchanger. The transition exhibits very many years of gross injustice, of bitter suffering, of absurd and ineffectual violations of the natural rights of man ; and of struggles between the capitalist and the labourer, for exclusive advantages, perpetuated by ignorant lawgivers, who could not see that the interest of all classes of producers is one and the same. We may not improperly devote a little space to the description of this dark and evil period. We shall see that while such a struggle goes forward—that is, while security of property and freedom of industry are not held as the interchangeable rights of the capitalist and the labourer—there can be little production and less accumulation. Wherever positive slavery exists—wherever the labourers are utterly deprived of their property in their labour, and are compelled to dispose of it without retaining any part of the character of voluntary exchangers—there are found idleness, ignorance, and unskilfulness ; industry is enfeebled—the oppressor and the oppressed are both poor—there is no national accumulation. The existence of slavery amongst the nations of antiquity was a great impediment to their progress in the arts of life. The community, in such nations, was divided into a caste of nobles called citizens, and a caste of labourers called slaves. The Romans were rich, in the common sense of the word, because they plundered other nations; but they could not produce largely when the individual spirit to industry was wanting. The industry of the freemen was rapine: the slaves were the producers. No man will work willingly when he is to be utterly deprived of the power of disposing at his own will of the fruits of his labour; no man will work skilfully when the same scanty pittance is doled out to each and all, whatever be the difference in their talents and knowledge. Wherever the freedom of industry is thus violated, property cannot be secure. If Rome had encouraged free labourers, instead of breeding menial slaves, it could not have happened that the thieves, who were constantly hovering round the suburbs of the city, like vultures looking out for carrion, should have been so numerous that, during the insurrection of Catiline, they formed a large accession to his army. But Rome had to encounter a worse evil than that of the swarms of highwaymen, who were ready to plunder whatever had been produced. Production itself was so feeble when carried on by the labour of slaves, that Columella, a writer on rural affairs, says the crops continued so gradually to fall off that there was a general opinion that the earth was growing old and losing its power of productiveness. Wherever slavery exists at the present day, there we find feeble production and national weakness. Poland, the most prolific corn country in Europe, is unquestionably the poorest country. Poland has been partitioned, over and over again, by governments that knew her weakness; and she has been said to have fallen "without a crime." That is not correct. Her "crime" was, and is, the slavery of her labourers. There is no powerful class between the noble and the serf or slave; and whilst this state of things endures, Poland can never be independent, because she can never be industrious, and therefore never wealthy.

England, as we have said, once groaned under the evils of positive slavery. The Anglo-Saxons had what they called "live money," such as sheep and slaves. To this cause may be doubtless attributed the easy conquest

of the country by the Norman invaders, and the oppression that succeeded that conquest. If the people had
been free, no king could have swept away the entire
population of a hundred thousand souls that dwelt in the
country between the Humber and the Tees, and converted a district of sixty miles in length into a dreary
desert, which remained for years without houses and
without inhabitants. This the Conqueror did. In the
reign of Henry II. the slaves of England were exported
in large numbers to Ireland. These slaves, or villeins,
as is the case in Russia and Poland at the present day,
differed in the degree of the oppression which was exercised towards them. Some, called "villeins in gross,"
were at the absolute disposal of the lord—transferable
from one owner to another, like a horse or a cow.
Others, called "villeins regardant," were annexed to
particular estates, and were called upon to perform whatever agricultural offices the lord should demand from
them, not having the power of acquiring any property,
and their only privilege being that they were irremove-
able except with their own consent. These distinctions
are not of much consequence, for, by a happy combination of circumstances, the bondmen of every kind, in the
course of a century or two after the Conquest, were
rapidly passing into the condition of free labourers. But
still capital was accumulated so slowly, and labour was so
unproductive, that the land did not produce the tenth
part of a modern crop; and the country was constantly
exposed to the severest inflictions of famine, whenever a
worse than usual harvest occurred.

In the reign of Edward III. the woollen manufacture
was introduced into England. It was at first carried on
exclusively by foreigners; but as the trade extended,
new hands were wanting, and the bondmen of the villages began to run away from their masters, and take
refuge in the towns. If the slave could conceal himself
successfully from the pursuit of his lord for a year and a
day, he was held free for ever. The constant attraction
of the bondmen to the towns, where they could work for
hire, gradually emboldened those who remained as culti-

vators to assert their own natural rights. The nobility complained that the villeins refused to perform their accustomed services; and that corn remained uncut upon the ground. At length, in 1351, the 25th year of Edward III., the class of free labourers was first recognised by the legislature; and a statute was passed, oppressive indeed, and impolitic, but distinctly acknowledging the right of the labourer to assume the character of a free exchanger. Slavery, in England, was not wholly abolished by statute till the time of Charles II.: it was attempted in vain to be abolished in 1526. As late as the year 1775, the colliers of Scotland were accounted *ascripti glebæ*—that is, as belonging to the estate or colliery where they were born and continued to work. It is not necessary for us further to notice the existence of villeinage or slavery in these kingdoms. Our business is with the slow progress of the establishment of the rights of free labourers—and this principally to show that, during the long period when a contest was going forward between the capitalists and the labourers, industry was comparatively unproductive. It was not so unproductive, indeed, as during the period of absolute slavery; but as long as any man was compelled to work, or to continue at work, or to receive a fixed price, or to remain in one place, or to follow one employment, labour could not be held to be free—property could not be held to be secure—capital and labour could not have cordially united for production—accumulation could not have been certain and rapid.

In the year 1349 there was a dreadful pestilence in England, which swept off large numbers of the people. Those of the labourers that remained, following the natural course of the great principle of demand and supply, refused to serve, unless they were paid double the wages which they had received five years before. Then came the "Statute of Labourers," of 1351, to regulate wages; and this statute enacted what should be paid to haymakers, and reapers, and thrashers; to carpenters, and masons, and tilers, and plasterers. No person was to quit his own village, if he could get work at these wages;

and labourers and artificers flying from one district to another, in consequence of these regulations, were to be imprisoned.

Good laws, it has been said, execute themselves. When legislators make bad laws, there requires a constant increase of vigilance and severity, and constant attempts at reconciling impossibilities, to allow such laws to work at all. In 1360, the statute of labourers was confirmed with new penalties, such as burning in the forehead with the letter F those workmen who left their usual abodes. Having controlled the wages of industry, the next step was for these blind lawgivers to determine how the workmen should spend their scanty pittance; and accordingly, in 1363, a statute was passed to compel workmen and all persons not worth forty shillings, to wear the coarsest cloth called russet, and to be served once a day with meat, or fish, and the offal of other victuals. We were not without imitations of such absurdities in other nations. An ordinance of the King of France, in 1461, determined that good and fat meat should be sold to the rich, while the poor should be allowed only to buy the lean and stinking.

While the wages of labour were fixed by statute, the price of wheat was constantly undergoing the most extraordinary fluctuations, ranging from 2s. a quarter, to 1l. 6s. 8d. It was perfectly impossible that any profitable industry could go forward in the face of such unjust and ridiculous laws. In 1376 the Commons complained that masters were *obliged* to give their servants higher wages to prevent their running away; and that the country was covered with *staf-strikers* and *sturdy rogues* who robbed in every direction. The villages were deserted by the labourers resorting to the towns, where commerce knew how to evade the destroying regulations of the statutes; and to prevent the total decay of agriculture, labourers were not allowed to move from place to place without letters patent:—any labourer, not producing such a letter, was to be imprisoned and put in the stocks. If a lad had been brought up to the plough till he was twelve years of age, he was compelled to continue in husbandry all his life; and in 1406 it was enacted that all children of parents not possessed of land should be brought up in the occupation of their parents. While the legislature, however, was passing these abominable laws, it was most effectually preparing the means for their extermination. Children were allowed to be sent to school in any part of the kingdom. When the light of education dawned upon the people, they could not long remain in the " darkness visible" that succeeded the night of slavery.

When the industry of the country was nearly annihilated by the laws regulating wages, it was found out that something like a balance should be preserved between wages and prices; and the magistrates were therefore empowered twice a year to make proclamation, according to the price of provisions, how much every workman should receive. The system, however, would not work well. In 1496 a new statute of wages was passed, the preamble of which recited that the former statutes had not been executed, because " the remedy by the said statutes is not very perfect." Then came a new remedy : that is, a new scale of wages for all trades ; regulations for the hours of work and of rest ; and penalties to prevent labour being transported from one district to another. As a necessary consequence of a fixed scale for wages, came another fixed scale for regulating the prices of provisions ; till at last, in the reign of Henry VIII., lawgivers began to open their eyes to the folly of their proceedings, and the preamble of a statute says " that dearth, scarcity, good cheap, and plenty of cheese, butter, capons, hens, chickens, and other victuals necessary for man's sustenance, happeneth, riseth, and chanceth, of so many and divers occasions, that it is very hard and difficile to put any certain prices to any such things." Yet they went on with new scales, in spite of the hardness of the task ; till at last some of the worst of these absurd laws were swept from the statute-book. The justices, whose principal occupation was to balance the scale of wages and labour, complained incessantly of the difficulty of the attempt ; and the statute of the 5th Elizabeth acknowledged that these old laws " could not be carried into execution without the great grief and burden of the poor labourer and hired man." Still new laws were enacted to prevent the freedom of industry working out plenty for capitalists as well as labourers; and at length, in 1601, a general assessment was directed for the support of the impotent poor, and for setting the unemployed poor to work. The capitalists at length paid a grievous penalty for their two centuries of oppression; and had to maintain a host of paupers, that had gradually filled the land during these unnatural contests. It would be perhaps incorrect to say, that these contests alone produced the paupers that required this legislative protection in the reign of Elizabeth; but certainly the number of those paupers would have been far less, if the laws of industry had taken their healthy and natural course,—if capital and labour had gone hand in hand to produce abundance for all, and fairly to distribute that abundance in the form of profits and wages, justly balanced by the steady operation of demand and supply in a free and extensive market.

The whole of these absurd and iniquitous laws, which had succeeded the more wicked laws of absolute slavery, proceeded from a struggle on the part of the capitalists in land against the growing power and energy of free labour. If the capitalists had rightly understood their interests, they would have seen that the increased production of a thriving and happy peasantry would have amply compensated them for all the increase of wages to which they were compelled to submit; and that at every step by which the condition of their labourers was improved their own condition was also improved. If then capital had worked naturally and honestly for the encouragement of labour, there would have been no lack of labourers; and it would not have been necessary to pass laws to compel artificers, under the penalty of the stocks, to assist in getting in the harvest (5 Eliz.). If the labourers in agriculture had been adequately paid, they would not have fled to the towns; and if they had not been liable to cruel punishments for the exercise of this their natural right, the country would not have been covered with " valiant rogues and sturdy beggars."

CHAPTER VIII.

It will be desirable to exhibit something like an average view of the extent of the possessions of all classes of society, and especially of the middling and labouring classes, in this country, at a period when the mutual rights of capitalists and labourers were so little understood as in the fourteenth century. We have shown how, at that time, there was a general round of oppression, resulting from ignorance of the proper interests of the productive classes; and it would be well also to show that during this period of disunion and contest between capital and labour, each plundering the other, and both plundered by arbitrary power, whether of the nobles or the crown, production went on very slowly and imperfectly, and that there was little to plunder and less to exchange. It is difficult to find the materials for such an inquiry. There is no very accurate record of the condition of the various classes of society before the invention of printing; and even after that invention we must be content to form our conclusions from a few scattered facts not recorded for any such purpose as we have in view, but to be gathered incidentally from slight observations which have come down to us. Yet enough remains to enable us to form a picture of tolerable accuracy; and in some points to establish conclusions which cannot be disputed. It is in the same way that our knowledge of the former state of the physical world must be derived from relics of that former state, to which the inquiries and comparisons of the present times have given an historical value. We know, for instance, that the animals of the southern countries once abounded in these islands, because we occasionally find their bones in quantities which could not have been accumulated unless such animals had been once native to these parts; and the remains of sea-shells upon the tops of hills now under the plough show us that even these heights have been heaved up from the bosom of the ocean. In the same way, although we have no complete picture of the state of property at the period to which we allude, we have evidence enough to describe that state from records which may be applied to this end, although preserved for a very different object.

In the reign of Edward III., Colchester, in Essex, was considered the tenth city in England in point of population. It then paid a poll-tax for 2955 lay persons. In 1301, about half a century before, the number of inhabitant housekeepers was 390; and the whole household furniture, utensils, clothes, money, cattle, corn, and every other property found in the town, was valued at 518*l*. 16*s*. 0¾*d*. This valuation took place on occasion of a subsidy or tax to the crown, to carry on a war against France; and the particulars, which are preserved in the Rolls of Parliament, exhibit with great minuteness the classes of persons then inhabiting that town, and the sort of property which each respectively possessed. The trades exercised in Colchester were the following:— baker, barber, blacksmith, bowyer, brewer, butcher, carpenter, carter, cobler, cook, dyer, fisherman, fuller, furrier, girdler, glass-seller, glover, linen-draper, mercer and spice-seller, miller, mustard and vinegar seller, old clothes seller, saddler, tailor, tanner, tyler, weaver, woodcutter, and wool-comber. If we look at a small town of the present day, where such a variety of occupations are carried on, we shall find that each tradesman has a considerable stock of commodities, abundance of furniture and utensils, clothes in plenty, some plate, books, and many articles of convenience and luxury to which the most wealthy dealers and mechanics of Colchester of the fourteenth century were utter strangers. That many places at that time were much poorer than Colchester there can be no doubt; for here we see the division of labour was pretty extensive, and that is always a proof that production is going forward, however imperfectly. We see, too, that the tradesmen were connected with manufactures in the ordinary use of the term; or there would not have been the dyer, the glover, the linen-draper, the tanner, the weaver, and the wool-comber. There must have been a demand for articles of foreign commerce, too, in this town, or we should not have had the spice-seller. Yet, with all these various occupations, indicating considerable profitable industry when compared with earlier stages in the history of this country, the whole stock of the town was valued at little more than 500l. Nor let it be supposed that this smallness of capital can be accounted for by the difference in the standard of money; although that difference is considerable. We may indeed satisfy ourselves of the small extent of the capital of individuals at that day, by referring to the inventory of the articles upon which the tax we have mentioned was laid at Colchester.

The whole stock of a carpenter's tools was valued at one shilling. They altogether consisted of two broad axes, an adze, a square, and a navegor or spoke-shave. Rough work must the carpenter have been able to perform with these humble instruments; but then let it be remembered that there was little capital to pay him for finer work, and that very little fine work was consequently required. The three hundred and ninety house-keepers of Colchester then lived in mud huts, with a rough door and no chimney. Harrison, speaking of the manners of a century later than the period we are describing, says, "There were very few chimneys even in capital towns: the fire was laid to the wall, and the smoke issued out at the roof, or door, or window. The houses were wattled, and plastered over with clay; and all the furniture and utensils were of wood. The people slept on straw pallets, with a log of wood for a pillow." When this old historian wrote, he mentions the erection of chimneys as a modern luxury. We had improved little upon our Anglo-Saxon ancestors in the article of chim-

neys. In their time Alcuin, an abbot who had ten
thousand vassals, writes to the emperor at Rome that he
preferred living in his smoky house to visiting the palaces
of Italy. This was in the ninth century. Five hundred
years had made little difference in the chimneys of Colchester. The nobility had hangings against the walls to
keep out the wind, which crept in through the crevices
which the builder's bungling art had left: the middle
orders had no hangings. Shakspere alludes to this rough
building of houses even in his time :—

> "Imperial Cæsar, dead and turned to clay,
> Might stop a hole to keep the wind away."

Even the nobility went without glass to their windows in
the fourteenth and fifteenth centuries. "Of old time,"
says Harrison, "our country-houses, instead of glass, did
use much lattice, and that made either of wicker or fine
rifts of oak, in checkerwise." When glass was introduced, it was for a long time so scarce that at Alnwick
Castle, in 1567, the glass was ordered to be taken out of
the windows, and laid up in safety, when the lord was
absent.

The mercer's stock-in-trade at Colchester was much
upon a level with the carpenter's tools. It was somewhat various, but very limited in quantity. The whole
comprised a piece of woollen cloth, some silk and fine
linen, flannel, silk purses, gloves, girdles, leather purses,
and needlework; and it was altogether valued at 3l.
There appears to have been another dealer in cloth and
linen in the town, whose store was equally scanty. We
were not much improved in the use of linen a century
later. We learn from the Earl of Northumberland's
household book, whose family was large enough to consume one hundred and sixty gallons of mustard during
the winter with their salt meat, that only seventy ells of
linen were allowed for a year's consumption. In the
fourteenth century none but the clergy and nobility wore
white linen. As industry increased, and the cleanliness
of the middle classes increased with it, the use of white
linen became more general; but, even at the end of the

next century, when printing was invented, the paper-makers had the greatest difficulty in procuring rags for their manufacture; and so careful were the people of every class to preserve their linen, that night-clothes were never worn. Linen was so dear that Shakspere makes Falstaff's shirts eight shillings an ell. The more sumptuous articles of a mercer's stock were treasured in rich families from generation to generation; and even the wives of the nobility did not disdain to mention in their wills a particular article of clothing, which they left to the use of a daughter or a friend. The solitary old coat of a baker came into the Colchester valuation; nor is this to be wondered at, when we find that even the soldiers at the battle of Bannockburn, about this time, were described by an old rhymer as " well near all naked."

The household furniture found in use amongst the families of Colchester consisted, in the more wealthy, of an occasional bed, a brass pot, a brass cup, a gridiron, and a rug or two, and perhaps a towel. Of chairs and tables we hear nothing. We learn from the Chronicles of Brantôme, a French historian of these days, that even the nobility sat upon chests in which they kept their clothes and linen. Harrison, whose testimony we have already given to the poverty of these times, affirms, that if a man in seven years after marriage could purchase a flock bed, and a sack of chaff to rest his head upon, he thought himself as well lodged as the lord of the town, " who peradventure lay seldom on a bed entirely of feathers." An old tenure in England, before these times, binds the vassal to find straw even for the king's bed. The beds of flock, the few articles of furniture, the absence of chairs and tables, would have been of less consequence to the comfort and health of the people, if they had been clean; but cleanliness never exists without a certain possession of domestic conveniences. The people of England, in the days of which we are speaking, were not famed for their attention to this particular. Thomas à Becket was reputed extravagantly nice, because he had his parlour strewed every day with clean straw. As late

as the reign of Henry VIII., Erasmus, a celebrated
scholar of Holland, who visited England, complains that
the nastiness of the people was the cause of the frequent
plagues that destroyed them; and he says, " their floors
are commonly of clay, strewed with rushes, under which
lie unmolested a collection of beer, grease, fragments,
bones, spittle, excrements of dogs and cats, and of everything that is nauseous." The elder Scaliger, another
scholar who came to England, abuses the people for
giving him no convenience to wash his hands. Glass
vessels were scarce, and pottery was almost wholly unknown. The Earl of Northumberland, whom we have
mentioned, breakfasted on trenchers and dined on pewter.
While such universal slovenliness prevailed as Erasmus
has described, it is not likely that much attention was
generally paid to the cultivation of the mind. Before the
invention of printing, at the time of the valuation of Colchester, books in manuscript, from their extreme costli-
ness, could be purchased only by princes. The royal
library of Paris, in 1378, consisted of nine hundred and
nine volumes,—an extraordinary number. The same
library now comprises upwards of four hundred thousand
volumes. But it may fairly be assumed that, where one
book could be obtained in the fourteenth century by
persons of the working classes, four hundred thousand
may be as easily obtained now. Here then was a privation which existed five hundred years ago, which debarred
our ancestors from more profit and pleasure than the want
of beds, and chairs, and linen; and probably, if this
privation had continued, and men therefore had not cultivated their understandings, they would not have learnt
to give any really profitable direction to their labour, and
we should still have been as scantily supplied with furniture and clothes as the good people of Colchester of
whom we have been speaking.

Let us see what accumulated supply, or capital, of food
the inhabitants of England had five centuries ago. Possessions in cattle are the earliest riches of most countries.
We have seen that cattle was called " live money ;" and
it is supposed that the word capital, which means stock

generally, was derived from the Latin word " capita," or heads of beasts. The law-term " chattels " is also supposed to come from cattle. These circumstances show that cattle were the chief property of our ancestors. Vast herds of swine constituted the great provision for the support of the people; and these were principally fed, as they are even now in the New Forest, upon acorns and beech-mast. In Domesday Book, a valuation of the time of William the Conqueror, it is always mentioned how many hogs each estate can maintain. Hume, the historian, in his Essays, alluding to the great herds of swine described by Polybius as existing in Italy and Greece, concludes that the country was thinly peopled and badly cultivated; and there can be no doubt that the same argument may be applied to England in the fourteenth century, although many swine were maintained in forests preserved for fuel. The hogs wandered about the country in a half-wild state, destroying, probably, more than they profitably consumed; and they were badly fed, if we may judge from a statute of 1402, which alleges the great decrease of fish in the Thames and other rivers, by the practice of feeding hogs with the fry caught at the wears. The hogs' flesh of England was constantly salted for the winter's food. The people had little fodder for cattle in the winter, and therefore they only tasted fresh meat in the summer season. The mustard and vinegar seller formed a business at Colchester, to furnish a relish for the pork. Stocks of salted meat are mentioned in the inventory of many houses there, and live hogs as commonly. But salted flesh is not food to be eaten constantly, and with little vegetable food, without severe injury to the health. In the early part of the reign of Henry VIII., not a cabbage, carrot, turnip, or other edible root, grew in England. Two or three centuries before, certainly, the monasteries had gardens with a variety of vegetables; but nearly all the gardens of the laity were destroyed in the wars between the houses of York and Lancaster. Harrison speaks of wheaten bread as being chiefly used by the gentry for their own tables; and adds, that the artificer and labourer

are " driven to content themselves with horse-corn, beans,
peason, oats, tares, and lentils.   There is no doubt that
the average duration of human life was at that period not
one-half as long as at the present day. The constant use
of salted meat, with little or no vegetable addition, doubtless contributed to the shortening of life, to say nothing
of the large numbers constantly swept away by pestilence
and famine.   Till lemon-juice was used as a remedy for
scurvy amongst our seamen, who also are compelled to
eat salted meat without green vegetables, the destruction
of life in the navy was something incredible.   Admiral
Hosier buried his ship's companies twice during a West
India voyage in 1726, partly from the unhealthiness of
the Spanish coast, but chiefly from the ravages of scurvy.
Bad food and want of cleanliness swept away the people
of the middle ages, by ravages upon their health that the
limited medical skill of those days could never resist.
Matthew Paris, an historian of that period, states that
there were in his time twenty thousand hospitals for
lepers in Europe.

   The slow accumulation of capital in the early stages of
the civilization of a country is in a great measure caused
by the indisposition of the people to unite for a common
good in public works, and the inability of governments
to carry on these works, when their principal concern is
war, foreign or domestic. The foundations of the civilization of this country were probably laid by our Roman
conquerors, who carried roads through the island, and
taught us how to cultivate our soil. Yet improvement
went on so slowly that, even a hundred years after the
Romans were settled here, the whole country was described as marshy. For centuries after the Romans made
the Watling-street and a few other roads, one district
was separated from another by the general want of these
great means of communication. Bracton, a law-writer of
the period we have been so constantly mentioning, holds
that if a man being at Oxford engage to pay money the
same day in London, he shall be discharged of his contract, as he undertakes a physical impossibility. We find,
as late as the time of Elizabeth, that her Majesty would

not stay to breakfast at Cambridge because she had to travel twelve miles before she could come to the place, Hinchinbrook, where she desired to sleep. Where there were no roads, there could be few or no markets. An act of parliament of 1272 says that the religious houses should not be compelled to *sell* their provisions—a proof that there were no considerable stores except in the religious houses. The difficulty of navigation was so great, that William Longsword, son of Henry II., returning from France, was during three months tossed upon the sea before he could make a port in Cornwall. Looking, therefore, to the want of commerce proceeding from the want of communication—looking to the small stock of property accumulated to support labour—and looking, as we have previously done, to the incessant contests between the small capital and the misdirected labour, both feeble, because they worked without skill—we cannot be surprised that the poverty of which we have exhibited a faint picture should have endured for several centuries, and that the industry of our forefathers must have had a long and painful struggle before it could have bequeathed to us such magnificent accumulations as we now enjoy.

CHAPTER IX.

THE writers who lived at the periods when Europe was slowly emerging from ignorance and poverty, through the first slight union of capital and labour as voluntary exchangers, complain of the increase of comforts as indications of the growing luxury and effeminacy of the people. Harrison says, " In times past men were contented to dwell in houses builded of sallow, willow, plum-tree, or elm ; so that the use of oak was dedicated to churches, religious houses, princes' palaces, noblemen's lodgings, and navigation. But now these are rejected, and nothing but oak any whit regarded. And yet see the change ; for when our houses were builded of willow, then had we oaken men ; but now that our houses are made of oak, our men are not only become willow, but many, through Persian delicacy crept in among us, altogether of straw, which is a sore alteration. In those days, the courage of the owner was a sufficient defence to keep the house in safety ; but now the assurance of the timber, double doors, locks, and bolts must defend the man from robbing. Now have we many chimneys, and our tenderlings complain of rheums, catarrhs, and poses. Then had we none but rere-dosses, and our heads did never ache." These complaints go upon the same principle that made it a merit in Epictetus, the Greek philosopher, to have had no door to his hovel. We think he would have been a wiser man if he had contrived to have had a door. A story is told of a Highland chief, Sir Evan Cameron, that himself and a party of his followers being benighted, and compelled to sleep

in the open air, when his son rolled up a ball of snow
and laid his head upon it for a pillow, the rough old man
kicked it away, exclaiming, " What, sir, are you turning
effeminate!" We doubt whether Sir Evan Cameron
and his men were braver than the English officers who
fought at Waterloo; and yet many of these marched from
the ball-room at Brussels in their holiday attire, and won
the battle in silk stockings. It is an old notion that
plenty of the necessaries and conveniences of life renders
a nation feeble. We are told that the Carthaginian soldiers which Hannibal carried into Italy were suddenly
rendered effeminate by the abundance which they found
around them at Capua. The commissariat of modern
nations goes upon another principle; and believes that
unless the soldier has plenty of food and clothing he will
not fight with alacrity and steadiness. The half-starved
soldiers of Henry V. won the battle of Agincourt; but
it was not because they were half-starved, but because
they roused their native courage to cut their way out of
the peril by which they were surrounded. When we
hear of ancient nations being enervated by abundance, we
may be sure that the abundance was almost entirely devoured by a few tyrants, and that the bulk of the people
were rendered weak by the destitution which resulted
from the unnatural distribution of riches. We read of
the luxury of the court of Persia — the pomp of the
seraglios, and of the palaces—the lights, the music, the
dancing, the perfumes, the silks, the gold, and the dia-
monds. The people are held to be effeminate. The
Russians, from the hardy north, can lay the Persian
monarchy any day at their feet. Is this national weakness caused by the excess of production amongst the
people, giving them so extravagant a command over the
necessaries and luxuries of life that they have nothing to
do but drink of the full cup of enjoyment? Mr. Fraser,
an English traveller, thus describes the appearance of
a part of the country which he visited in 1821 :—" The
plain of Yezid-Khaust presented a truly lamentable picture of the general decline of prosperity in Persia. Ruins
of large villages thickly scattered about with the skeleton-

like walls of caravansaries and gardens, all telling of
better times, stood like *memento moris* (remembrances of
death) to kingdoms and governments; and the whole
plain was dotted over with small mounds, which indicate
the course of cannauts (artificial streams for watering
the soil), once the source of riches and fertility, now
all choked up and dry; for there is neither man nor cultivation to require their aid." Was it the luxury of the
people which produced this decay—the increase of their
means of production—their advancement in skill and capital; or some external cause which repressed production, and destroyed accumulation both of outward wealth
and knowledge? "Such is the character of their rulers,"
says Mr. Fraser, "that the only measure of their demands is the power to extort on one hand, and the ability
to give or retain on the other." Where such a system
prevails, all accumulated labour is concealed, for it would
otherwise be plundered. It does not freely and openly
work to encourage new labour. Burckhardt, the traveller
of Nubia, saw a farmer who had been plundered of everything by the pacha, because it came to the ears of the
savage ruler that the unhappy man was in the habit of
eating wheaten bread; and that, he thought, was too
great a luxury for a subject. If such oppressions had not
long ago been put down in England, we should still have
been in the state of Colchester in the fourteenth century.
When these iniquities prevailed, and there was neither
freedom of industry nor security of property — when
capital and labour were not united—when all men consequently worked unprofitably, because they worked without division of labour, accumulation of knowledge, and
union of forces — there was universal poverty, because
there was feeble production. Slow and painful were the
steps which capital and labour had to make before they
could emerge, even in part, from this feeble and degraded state. But that they have made a wonderful
advance in five hundred years will not be difficult to
show.

Let us first compare the Colchester of the nineteenth
century with the Colchester of the fourteenth, in a few
particulars.

In the reign of Edward III. Colchester numbered three hundred and fifty-nine houses of mud, without chimneys, and with latticed windows. In the reign of Queen Victoria it has seven hundred and forty-four houses, each at a rent above ten pounds. The houses below ten pounds are not mentioned in the return from which we derive this information. Houses of ten pounds a-year and upwards are commonly built of brick, and slated or tiled; secured against wind and weather; with glazed windows and with chimneys; and generally well ventilated. The worst of these houses are supplied, as fixtures, with a great number of conveniences, such as grates, and cupboards, and fastenings. To many of such houses gardens are attached, wherein are raised vegetables and fruits that kings could not command two centuries ago. Houses such as these are composed of several rooms—not of one room only, where the people are compelled to eat and sleep and perform every office, perhaps in company with pigs and cattle—but of a kitchen, and often a parlour, and several bed-rooms. These rooms are furnished with tables, and chairs, and beds, and cooking-utensils. There is ordinarily, too, something for ornament and something for instruction;—a piece or two of china, silver spoons, books, and not unfrequently a watch or clock. The useful pottery is abundant and of really elegant forms and colours; drinking-vessels of glass are not uncommon. The inhabitants are not scantily supplied with clothes. The females are decently dressed, having a constant change of linen, and gowns of various patterns and degrees of fineness. Some, even of the humbler classes, are not thought to exceed the proper appearance of their station if they wear silk. The men have decent working habits, strong shoes and hats,.and a respectable suit for Sundays, of cloth often as good as is worn by the highest in the land. Every one is clean; for no house above the few hovels which still deform the country is without soap and bowls for washing, and it is the business of the females to take care that the linen of the family is constantly washed. The children, almost universally, receive instruction in some public establishment; and when the labour of the day is over, the father thinks the time unprofitably spent unless he burns a candle to enable him to read a book or the newspaper. The food which is ordinarily consumed is of the best quality. Wheaten bread is no longer confined to the rich; animal food is not necessarily salted, and salt meat is used principally as a variety; vegetables of many sorts are plenteous in every market, and these by a succession of care are brought to higher perfection than in the countries of more genial climate from which we have imported them; the productions too of distant regions, such as spices, and coffee and tea, are universally consumed almost by the humblest in the land. Fuel, also, of the best quality is abundant and comparatively cheap.

If we look at the public conveniences of a modern English town, we shall find the same striking contrast. Water is brought not only into every street, but into every house; the dust and dirt of a family is regularly removed without bustle or unpleasantness; the streets are paved, and lighted at night; roads in the highest state of excellence connect the town with the whole kingdom, and by means of railroads a man can travel several hundred miles in a few hours, and more readily than he could ten miles in the old time; and canal and sea navigation transport the weightiest goods with the greatest facility from each district to the other, and from each town to the other, so that all are enabled to apply their industry to what is most profitable for each and all. Every man, therefore, may satisfy his wants, according to his means, at the least possible expense of the transport of commodities. Every tradesman has a stock ready to meet the demand;—and thus the stock of a very moderately wealthy tradesman of the Colchester of the present day is worth more than all the stock of all the different trades that were carried on in the same place in the fourteenth century. To effect these public conveniences, millions of capital have been invested, which sums have afforded profitable labour to millions of workmen. Look at the iron trade, which has so large a share in all public works. In the year 1788, sixty thousand tons of cast-iron were manufactured. In the year 1843, the amount of the produce of cast-iron was one million five hundred thousand tons. A large portion of this enormous increase has been applied to the internal improvement of the country, in water-pipes, gas-pipes, bridges, railroads.

But to allow us to form a tolerable estimate of the increased production and accumulation of this country, we must take a few general points of comparison, which may enable us to estimate the astonishing extent of this production and accumulation, more accurately even than from the individual case we have exhibited.

And first, of the Population of the country;—for an increase of population always shows an increase of production, for without increased production the amount of population must remain stationary; with diminished production it must become less; and if there were no production and therefore no accumulation, population would be altogether extinguished. Mr. Turner, the historian of the Anglo-Saxons, has estimated from Domesday Book, that the population of England at the time of the Norman conquest somewhat exceeded two millions. It has been estimated by Mr. Chalmers that in 1377 the population did not exceed 2,350,000 souls. There was an increase, therefore, of only the third of a million in three centuries and a half. From 1377 to 1841, a period of little more than four centuries and a half, the population of England had increased to nearly fifteen millions, or more than six times the amount of the population of 1377. The increased production of the country must have gone forward in the same proportion—to say nothing of the much greater comparative increase of production demanded by the change for the better in the habits of every class of the consumers. We have no materials for comparing the general production of five hundred years ago with the general production of the present day;—yet every man may compare in his own mind the state in which he himself lives, and the state in which the people of Colchester lived at the time we have described in the last chapter. To assist this comparison we will furnish a few

E

particulars of the present home consumption of the kingdom, in the great staple articles of her commerce and manufactures. We only take those articles which can be accurately estimated.

Of *wheat*, fifteen million quarters are annually consumed in the United Kingdom. Of *malt*, forty million bushels are annually used in breweries and distilleries in the United Kingdom; and there are forty-six thousand acres under cultivation with *hops*. Of the quantity of potatoes and other vegetables consumed we have no accounts. Of meat, about one million six hundred thousand head of cattle and sheep are sold during the year in Smithfield market alone, which is probably about a tenth of the consumption of the whole kingdom. The quantity of *tea* consumed in the United Kingdom is about thirty-seven million lbs. annually. Of *sugar* rather more than four million hundred-weights, or about five hundred million lbs. every year,—which is a consumption of about eighteen lbs. for every individual, reckoning the population at twenty-seven millions; and of *coffee* about twenty-seven million lbs. are annually consumed. Of *soap*, one hundred and seventy million lbs. are consumed. Of sea-borne *coals* alone there are seven million five hundred thousand tons consumed, and the total consumption of coal in Great Britain is estimated at thirty million tons. Of clothing, we annually manufacture above four hundred million lbs. of cotton wool, which produce above fourteen hundred million yards of calico and various other cotton fabrics, and of these we export about half; so that above seven hundred million yards remain for home consumption, being about twenty-six yards annually for each person in a population of twenty-seven millions. The woollen manufacture consumes about forty million lbs. of foreign wool, besides almost the whole of the wool produced at home, estimated at about a hundred and eighty millions of lbs. Of *hides* and *skins* above fifty million are annually tanned and dressed. Of *paper*, nearly one hundred million lbs. are yearly manufactured, which is about four million reams, of five hundred sheets to the ream.

To carry on the commerce of this country with foreign nations and between distant parts of the United Kingdom, there are twenty-three thousand ships in constant employ, belonging to our own merchants. To carry on the commerce with ourselves, the total length of our turnpike-roads (in England and Wales) is twenty thousand miles, besides nearly one hundred thousand miles of other highways, three thousand miles of canals, and about two thousand five hundred miles of railroads, the latter constructed at a cost of about seventy million pounds sterling. To produce food for the inhabitants of the country we have about forty-five million acres under cultivation. To clothe them we have millions of spindles worked by steam, instead of a few thousands turned by hand as they were a century ago. The incomes of persons who have 150*l.* a-year and upwards amount to an aggregate of about one hundred and eighty million pounds sterling per annum ; of which rather more than one-half is derived from real property. The fixed capital of the country insured in fire-offices, that insurance being far short of its real amount, is above six hundred and eighty million pounds sterling. The fixed capital uninsured, or not represented by this species of insurance, is perhaps as much. The capital expended in improvement in land, is, we should conceive, equal to the capital which is represented by houses, and furniture, and shipping, and stocks of goods. The public capital of the country expended in roads, canals, docks, harbours, and buildings, is equal to at least half the private capital. All this capital is the accumulated labour of two thousand years, when the civilization of the country first began. The greater portion of it is the accumulated labour of the last four hundred years, when labour and capital, through the partial abolition of slavery, first began to work together with freedom, and therefore with energy and skill.

CHAPTER X.

Two of the most terrific famines that are recorded in the
history of the world occurred in Egypt — a country
where there is greater production, with less labour, than
is probably exhibited in any other region. The principal labourer in Egypt is the river Nile, whose periodical
overflowings impart fertility to the thirsty soil, and produce in a few weeks that abundance which the labour of
the husbandman might not hope to command if employed
during the whole year. But the Nile is a workman that
cannot be controlled and directed, even by capital, the
great controller and director of all work. The influences
of heat, and light, and air, are pretty equal in the same
places. Where the climate is most genial, the cultivators
have least labour to perform in winning the earth; where
it is least genial the cultivators have most labour. The
increased labour balances the small natural productiveness.
But the inundation of a great river cannot be depended
upon like the light and heat of the sun. For two seasons
the Nile refused to rise, and labour was not prepared to
compensate for this refusal;—the ground refused to produce; the people were starved.

We mention these famines of Egypt to show that *certainty* is the most encouraging stimulus to every operation of human industry. We know that production as
invariably follows a right direction of labour, as day succeeds to night. We believe that it will be dark to-night
and light again to-morrow, because we know the general
laws which govern light and darkness, and because our
experience shows us that those laws are constant and

uniform.   We know that if we plough, and manure, and sow the ground, a crop will come in due time, varying indeed in quantity according to the season, but still so constant upon an average of years, that we are justified in applying large accumulations and considerable labour to the production of this crop.   It is this certainty that we have such a command of the productive powers of nature as will abundantly compensate us for the incessant labour of directing those forces, which has during a long course of industry heaped up the manifold accumulations which we described in the last chapter, and which enables production annually to go forward to the immense extent which we there exhibited.   The long succession of labour, which has covered this country with wealth, has been applied to the encouragement of the productive forces of nature, and the restraint of the destroying.   No one can doubt that the instant the labour of man ceases to direct those productive natural forces, the destroying forces immediately come into action. Take the most familiar instance—a cottage whose neat thatch was never broken, whose latticed windows were always entire, whose white-washed walls were ever clean, round whose porch the honeysuckle was trained in regulated luxuriance, whose garden bore nothing but what the owner planted.   Remove that owner.   Shut up the cottage for a year, and leave the garden to itself.   The thatched roof is torn off by the wind and devoured by mice, the windows are driven in by storms, the walls are soaked through with damp and are crumbling to ruin, the honeysuckle obstructs the entrance which it once adorned, the garden is covered with weeds which years of after-labour will have difficulty to destroy :

> " It was a plot
> Of garden-ground run wild, its matted weeds
> Mark'd with the steps of those, whom, as they passed,
> The gooseberry trees that shot in long lank slips,
> Or currants, hanging from their leafless stems
> In scanty strings, had tempted to o'erleap
> The broken wall."

Apply this principle upon a large scale. Let the productive energy of a country be suspended through some great cause which prevents its labour continuing in a profitable direction. Let it be overrun by a conqueror, or plundered by domestic tyranny of any kind, so that capital ceases to work with security. The fields suddenly become infertile, the towns lose their inhabitants, the roads grows to be impassable, the canals are choked up, the rivers break down their banks, the sea itself swallows up the land. Shakspere, a great political reasoner as well as a great poet, has described such effects, in that part of Henry V. when the Duke of Burgundy exhorts the rival kings to peace:—

> "Let it not disgrace me,
> If I demand, before this royal view,
> What rub, or what impediment, there is,
> Why that the naked, poor, and mangled peace,
> Dear nurse of arts, plenties, and joyful births,
> Should not, in this best garden of the world,
> Our fertile France, put up her lovely visage?
> Alas! she hath from France too long been chas'd;
> And all her husbandry doth lie on heaps,
> Corrupting in its own fertility.
> Her vine, the merry cheerer of the heart,
> Unpruned dies: her hedges even-pleached,
> Like prisoners wildly over-grown with hair
> Put forth disordered twigs: her fallow leas,
> The darnel, hemlock, and rank fumitory
> Doth root upon; while that the coulter rusts,
> That should deracinate such savagery:
> The even mead, that erst brought sweetly forth
> The freckled cowslip, burnet, and green clover,
> Wanting the scythe, all uncorrected, rank,
> Conceives by idleness; and nothing teems
> But hateful docks, rough thistles, kecksies, burs,
> Losing both beauty and utility:
> And as our vineyards, fallows, meads, and hedges,
> Defective in their natures, grow to wildness;
> Even so our houses, and ourselves, and children,
> Have lost, or do not learn, for want of time,
> The sciences that should become our country."

We have heard it said that Tenterden steeple was the cause of Goodwin Sands. The meaning of the saying is, that the capital which was appropriated to keep out the sea from that part of the Kentish coast was diverted to the building of Tenterden steeple; and there being no funds to keep out the sea, it washed over the land.* The Goodwin Sands remain to show that man must carry on a perpetual contest to keep in subjection the forces of nature, which, as is said of fire, one of the forces, are good servants but bad masters. But these examples show, also, that in the social state our control of the physical forces of nature depends upon the right control of our own moral forces. There was injustice, doubtless, in misappropriating the funds which restrained the sea from devouring the land. Till men know that they shall work with justice on every side, they work feebly and unprofitably. England did not begin to accumulate largely and rapidly till the rights both of the poor man and the rich were to a certain degree established—till industry was free, and property secure. Let any circumstances again arise which may be powerful enough to destroy, or even molest, the freedom of industry and the security of property, and we should work once more without certainty. The elements of prosperity would not be constant and uniform. We should work with the apprehension that some hurricane of tyranny, no matter from what power, would arise, which would sweep away accumulation. When that hurricane did not rise, we might have comparative abundance, like the people of Egypt during the inundation of the Nile. We then should have an inundation of tranquillity. But if the tranquillity were not present—if lawless violence stood in the place of justice and security—we should be like the people of Egypt when the Nile did not overflow. We should suffer the extremity of misery; and that possible extremity would produce an average misery, even if tranquillity did return, because security had not returned. We should, if this state of things long abided,

* Grey's Notes to Hudibras.

by degrees go back to the condition of Colchester in the fourteenth century, and thence to the universal marsh of two thousand years ago. The place where London stands would be, as it once was, a wilderness for howling wolves. The few that produced would again produce laboriously and painfully, without skill and without division of labour, because without accumulation; and it would probably take another thousand years, if men again saw the absolute need of security, to re-create what security has accumulated for our present use.

From the moment that the industry of this country began to work with security, and capital and labour applied themselves in union — perhaps not a perfect union, but still in union — to the great business of production, they worked with less and less expenditure of unprofitable labour. They continued to labour more and more profitably, as they laboured with knowledge. The labour of all rude nations, and of all uncultivated individuals, is labour with ignorance. Peter the wild boy, whom we have already mentioned, could never be made to perceive the right direction of labour, because he could not trace it through its circuitous courses for the production of utility. He would work under control, but, if left to himself, he would not work profitably. Having been trusted to fill a cart with manure, he laboured with diligence till the work was accomplished; but no one being at hand to direct him, he set to work as diligently to unload the cart again. He thought, as too many think even now, that the good was in the labour, and not in the results of the labour. The same ignorance exhibits itself in the unprofitable labour and unprofitable application of capital, even of persons far removed beyond the half-idiocy of Peter the wild boy. In the thirteenth century many of the provinces of France were overrun with rats, and the people, instead of vigorously hunting the rats, were persuaded to carry on a process against them in the ecclesiastical courts; and there, after the cause of the injured people and injuring rats was solemnly debated, the rats were declared cursed and excommunicated if they did not retire in six days. The historian does not add that the rats obeyed the injunction; and doubtless the farmers were less prepared to resort to the profitable labour of chasing them to death when they had paid the ecclesiastics for the unprofitable labour of their excommunication. There is a curious instance of unprofitable labour given in a book on the Coal Trade of Scotland, written as recently as 1812. The people of Edinburgh had a passion for buying their coals in immense lumps, and, to gratify this passion, the greatest care was taken not to break the coals in any of the operations of conveying them from the pit to the cellar of the consumer. A wall of coals was first built within the pit, another wall under the pit's mouth, another wall when they were raised from the pit, another wall in the waggon which conveyed them to the port where they were shipped, another wall in the hold of the ship, another wall in the cart which conveyed them to the consumer, and another wall in the consumer's cellar; and the result of these seven different buildings-up and takings-down was, that after the consumer had paid thirty per cent. more for these square masses of coal than for coal shovelled together in large and small pieces, his servant had daily to break the large coals to bits to enable him to make any use of them. It seems extraordinary that such waste of labour and capital should have existed amongst a highly acute and refined community, within the last thirty years. They, perhaps, thought they were making good for trade, and therefore submitted to the evil; while the Glasgow people, on the contrary, by saving thirty per cent. in their coals, had that thirty per cent. to bestow upon new enterprises of industry, and for new encouragements to labour.

The unprofitable applications of capital and labour which the early history of the civilization of every people has to record, and which, amongst many, have subsisted even whilst they held themselves at the height of refinement, have been fostered by the ignorance of the great, and even of the learned, as to the causes which, advancing production or retarding it, advanced or retarded their own interests, and the interests of all the community. Princes and statesmen, prelates and philosophers, were equally ignorant of

> " What makes a nation happy, and keeps it so;
> What ruins kingdoms, and lays cities flat."

It was enough for them to consume ; they thought it beneath them to observe even, much less to assist in, the direction of production. This was ignorance as gross as that of Peter the wild boy, or the excommunication of rats. It has always been the fashion of ignorant greatness to despise the mechanical arts. The pride of the Chinese mandarins was to let their nails grow as long as their fingers, to show that they never worked. Even European nobles once sought the same absurd distinction. In France, under the old monarchy, no descendant of a nobleman could embark in trade without the highest disgrace ; and the principle was so generally recognized as just, that a French writer, even as recently as 1758, reproaches the sons of the English nobility for the contrary practice, and asks, with an air of triumph, how can a man be fit to serve his country in Parliament after having meddled with such paltry concerns as those of commerce ? Montesquieu, a writer in most respects of enlarged views, holds that it is beneath the dignity of governments to interfere with such trumpery things as the regulation of weights and measures. Society might have well spared the interference of governments with weights and measures if they had been content to leave all commerce equally free. But, in truth, the regulation of weights and measures is almost a solitary exception to the great principle which governments ought to practise, of not interfering, or interfering little with commerce.

Louis XIV. did not waste more capital and labour by his ruinous wars, and by his covering France with fortifications and palaces, than by the perpetual interferences of himself and his predecessors with the freedom of trade, which compelled capital and labour to work unprofitably. The naturally slow progress of profitable industry is rendered more slow by the perpetual inclination of those in authority to divert industry from its natural and profitable channels. It was therefore wisely said by a committee of merchants to Colbert, the Prime Minister of France in the reign of Louis XIV., when he asked them what measures government could adopt to promote the interests of commerce,—" Let us alone, permit us quietly to manage our own business." It is undeniable that the interests of all are best promoted when each is left free to attend to his own interests, under the necessary social restraints which prevent him doing a positive injury to his neighbour. It is thus that agriculture and manufactures are essentially allied in their interests ; that unrestrained commerce is equally essential to the real and permanent interests of agriculture and manufactures ; that capital and labour are equally united in their interests, whether applied to agriculture, manufactures, or commerce ; that the producer and the consumer are equally united in their most essential interest, which is, that there should be cheap production. While these principles are not understood at all, and while they are imperfectly understood, as they still are by many classes and individuals, there must be a vast deal of unprofitable expenditure of capital, a vast deal of unprofitable labour, a vast deal of bickering and heart-burning between individuals who ought to be united, and classes who ought to be united, and nations who ought to be united ; and as long as it is not felt by all that their mutual rights are understood and will be respected, there is a feeling of insecurity which more or less affects the prosperity of all. The only remedy for these evils is the extension of knowledge. Louis XV. proclaimed to the French that the English were their "véritables ennemis," their true enemies. When knowledge is triumphant it will be found that there are no "véritables ennemis," either among nations, or classes, or individuals. The prejudices by which nations, classes, and individuals are led to believe that the interest of one is opposed to the interest of another, are, nine times out of ten, as utterly absurd as the reason which a Frenchman once gave for hating the English—which was, " that they poured melted butter on their roast veal ; " and this was not more ridiculous than the old denunciation of the English against the French, that " they ate frogs, and wore wooden shoes." When the world is disabused of the belief that the wealth of one nation, class, or individual must be created by the loss of another's wealth, then, and then only, will all men steadily and harmoniously apply to produce and to enjoy—to acquire prosperity and happiness—lifting themselves to the possession of good

" By Reason's light, on Resolution's wings."

CHAPTER XI.

ONE of the most striking and lamentable effects of the want of knowledge, producing disunions amongst mankind that are injurious to the interests of each and all, is the belief which still exists amongst many of the working men of these kingdoms, that the powers and arrangements which Capital has created and devised for the advancement of production are injurious to them in their character of producers. The great forces by which capital and labour now work,—forces which are gathering strength every day,—are accumulation of skill and division of employments. It will be for us to show that the applications of science to the manufacturing arts have the effect of ensuring cheap production and increased employment. These applications of science are principally displayed in the use of MACHINERY; and we shall endeavour to prove that, although individual labour may be partially displaced, or unsettled for a time, by the use of this cheaper and better power than unassisted manual labour, all are great gainers by the general use of that power. Through that power all principally possess, however poor they may be, many of the comforts which make the difference between man in a civilized and man in a savage state; and further, that in consequence of machinery having rendered productions of all sorts cheaper, and therefore caused them to be more universally purchased, it has really increased the demand for that manual labour, which it appears to some, reasoning only from a few instances, it has a tendency to diminish.

In the year 1827, a Committee of the House of Commons was appointed to examine into the subject of Emigration—that is, to see whether it was desirable and practicable to remove distressed labourers from the United Kingdom to distant places, where their labour might be profitably employed to themselves and others.   The first person examined before that Committee was Joseph Foster, a working weaver of Glasgow.   He told the Committee, that he and many others, who had formed themselves into a society, were in great distress ; that numbers of them worked at the *hand-loom* from eighteen to nineteen hours a-day, and that their earnings, at the utmost, did not amount to more than seven shillings a-week, and that sometimes they were as low as four shillings.   That twenty years before that time they could readily earn a pound a-week by the same industry ; and that as *power-loom* weaving had increased, the distress of the hand-weavers also had increased in the same proportion.   A power-loom is one worked by machinery, and not by the hand of man, as most of our readers perhaps know.   The Committee then put to Joseph Foster the following questions, and received the following answers :—

Q. " Are the Committee to understand that you attribute the insufficiency of your remuneration for your labour to the introduction of machinery ?

A. Yes.

Q. Do you consider, therefore, that the introduction of machinery is objectionable ?

A. We do not.   The weavers in general, of Glasgow and its vicinity, do not consider that machinery can or ought to be stopped, or put down.   They know perfectly well that machinery must go on, that it will go on, and that it is impossible to stop it.   They are aware that every implement of agriculture or manufacture is a portion of machinery, and, indeed, everything that goes beyond the teeth and nails (if I may use the expression) is a machine.   I am authorized, by the majority of our society, to say, that I speak their minds, as well as my own, in stating this."

The difference between those who object to machines,

and the persons who think with Joseph Foster, is, as it appears to us, a want of knowledge. We desire to communicate that knowledge. It would be out of our power to impart this knowledge *at all* without machinery : and, therefore, we shall begin by explaining how the machinery which gives *knowledge* of any sort by the means of books, is a vast blessing, when compared with slower methods of multiplying written language ; and how, by the aid of this machinery, we can produce a book without any limit in point of the number of copies, with great rapidity, and at a small price.

It is nearly four hundred years since the art of printing books was invented. Before that time all books were written by the hand. There were many persons employed to copy out books, but they were very dear, although the copiers had small wages. A Bible was sold for thirty pounds in the money of that day, which was equal to a great deal more of our money. Of course, very few people had Bibles or any other books. An ingenious man invented a mode of imitating the written books by cutting the letters on wood, and taking off copies from the wooden blocks by rubbing the sheet on the back ; and soon after other clever men thought of casting metal types or letters, which could be arranged in words, and sentences, and pages, and volumes ; and then a machine, called a printing-press, upon the principle of a screw, was made to stamp impressions of these types so arranged. There was an end, then, at once to the trade of the pen-and-ink copiers ; because the copiers in types, who could press off several hundred books while the writers were producing one, drove them out of the market. A single printer could do the work of at least two hundred writers. At first sight this seems a hardship, for a hundred and ninety-nine people might have been, and probably were, thrown out of their accustomed employment. But what was the consequence in a year or two ? Where one written book was sold a thousand printed books were required. The old books were multiplied in all countries, and new books were composed by men of talent and learning, because they could then find numerous readers. The printing press did the work more neatly and more correctly than the writer, and it did it infinitely cheaper. What then? The writers of books had to turn their hands to some other trade, it is true ; but type-founders, paper-makers, printers, and book-binders, were set to work, by the new art or machine, to at least a hundred times greater number of persons than the old way of making books employed. If the pen-and-ink copiers could break the printing-presses and melt down the types that are used in London alone at the present day, twenty thousand people would at least be thrown out of employment to make room for two hundred at the utmost ; and what would be even worse than all this misery, books could only be purchased, as before the invention of printing, by the few rich, instead of being the comforters and best friends of the millions who are now within reach of the benefits and enjoyments which they bestow.

The *cheapness of production* is the great point to which we shall call attention, as we give other examples of the good of machinery. In the case of books produced by the printing-press we have a cheap article, and an increased number of persons engaged in manufacturing that article. In almost all trades the introduction of machines has, sooner or later, the like effects. This we shall show as we go on. But to make the matter even more clear, we shall direct notice to the very book the reader holds in his hand, to complete our illustration of the advantages of machinery to the consumer, that is, to the person who wants and buys the article consumed, as well as to the producer, or the person who manufactures the article produced.

This little book is intended to consist of about 240 pages, to be printed, eighteen on a side, upon seven sheets of printing paper, called by the makers demy. These seven sheets of demy, at the price charged in the shops, would cost three-pence halfpenny. If the same number of words were written, instead of being printed —that is, if the closeness and regularity of printing were superseded by the looseness and unevenness of writing,

—they would cover 240 pages, or 60 sheets, of the paper called foolscap, which would cost in the shops three shillings; and we should offer a book difficult instead of easy to read, because writing is much harder to decipher than print. Here, then, besides the superiority of the workmanship, is at once a saving of two shillings and eight-pence halfpenny to the consumer, by the invention of printing, all other things being equal. But the great saving is to come. Work as hard as he could, a writer could not transcribe this little book upon these 240 pages of foolscap in less than ten days; and he would think himself very ill paid to receive thirty shillings for the operation. Adding, therefore, a profit for the publisher and retail tradesman, a single written copy of this little book, which is sold for a shilling, could not be produced for two pounds. Is it not perfectly clear, then, if there were no printing-press, if the art of printing did not exist, that if we found purchasers at all for this dear book at the cost of two pounds, we should only sell, at the utmost, a fortieth part of what we now sell; that instead of selling ten thousand copies we could only sell, even if there were the same quantity of book-buying funds amongst the few purchasers as amongst the many, two hundred and fifty copies; and that therefore, although we might employ two hundred and fifty writers for a week, instead of about twenty printers in the same period, we should have forty times less employment for paper-makers, ink-makers, book-binders, and many other persons, besides the printers themselves, who are called into activity by the large demand which follows cheapness of production.

It will be perceived, without having the subject dwelt upon, that if we could not publish this book *cheaply*, we could not publish it *extensively;* that, in fact, the book would be useless; that it would be a mere curiosity; that we should not attempt to multiply any copies, because those whose use it was intended for could not buy it. It is also perfectly clear, that if, by any unnatural reduction of the wages of labour, such as happens to the Hindoo, who works at weaving muslin for about sixpence a-week, we could get copiers to produce the book as cheaply as the printing-press (which is impossible), we could not send it to the world as *quickly*. We can get ten thousand copies of this book printed in a week, by the aid of about twelve compositors, and two printing machines, each machine requiring two boys and a man for its guidance. To transcribe ten thousand copies in the same time would require more than ten thousand penmen. Is it not perfectly evident, therefore, that if printing, which is a cheap and a rapid process, were once again superseded by writing, which is an expensive and a slow operation, neither this book, nor any other book, could be produced for the use of the people; that knowledge, upon which every hope of bettering their condition must ultimately rest, would again become the property of a very few; and that mankind would lose the greater part of that power, which has made, and is making them, truly independent, and which will make them virtuous and happy?

The same principle applies to any improvement of the machinery used in printing, or in the manufacture of the paper upon which books are printed. By the use of the printing-machine, instead of the printing-press (which machine is only profitably applicable to books printed in large numbers), the cost of production is diminished at least one-tenth; and by the use of the machine for making paper, a better article is produced, also at a lower rate. This book is printed upon paper as fine as is needful, instead of paper of a wretched quality; because the paper-machine has diminished the cost of production, by working up the pulp of which paper is composed more evenly, and therefore with a saving. And from both causes united, the diminished price of printing by the machine instead of printing by hand, and the diminished price of machine-made paper, the buyers of this book have seven sheets, instead of five sheets, for a shilling. Thus, not only is the price lessened to the consumer, by the increase of the quantity, but two-sevenths more paper, more ink, one-sixth more labour of the compositor, or printer who arranges the types, more labour of the sewer, or binder of the book; all these additions of direct labour, and of materials produced by labour, are consumed. In selling this book, therefore, for a shilling, we give two-sevenths more matter than you could have had without these new inventions; if we were to take away two-sevenths in quantity, we could lessen the price, and give the smaller book for ninepence. Thus, then, there is a decided advantage to the consumer in the diminished cost of the production, and an ample equivalent in mere labour (which, bear always in mind, is the means of producing commodities, and not the end for which they are produced), in the place of labour thrust out by the printing-machine and the paper-machine.

We think that in the article of *Books* we have proved that machinery has rendered productions cheaper, and has increased the demand for manual labour, and consequently the number of labourers; and that, therefore, machinery applied to books is not objectionable. We proceed from books to articles of actual necessity.

CHAPTER XII.

AMONGST the many accounts which the newspapers in December, 1830, gave of the destruction of machinery by agricultural labourers, we observed that in the neighbourhood of Aylesbury, a band of mistaken and unfortunate men destroyed all the machinery of many farms, *down even to the common drills.* The men conducted themselves, says the county newspaper, with civility; and such was their consideration, that they moved the machines out of the farm-yards, to prevent injury arising to the cattle from the nails and splinters that flew about while the machinery was being destroyed. They *could not make up their minds* as to the propriety of destroying a horse-churn, and therefore that machine was passed over.

We will suppose, by way of argument, that there were no laws to repress such outrages; that if the labourers in agriculture chose to believe that not only thrashing-machines were injurious to them, but that every one of those ingenious implements which have aided in rendering British agriculture the most perfect in the world, was equally harmful, they might, without interruption, break them to pieces. We will further suppose, that while these proceedings go forward, the landowner looks on, the farmer looks on, the magistrate does not stir, there are no judges or juries in the land, the people in the towns leave the machine-breakers to their own devices. So they break on—thrashing-machine, winnowing-machine, chaff-cutting-machine, drill, and every other new-fangled invention, as they call these things. We will suppose, still further, that the farmer yields to all this violence; that the violence has the effect which it was meant to have upon him; and that he takes on all the hands which were out of employ, to thrash and winnow, to cut chaff, to plant with a dibber instead of with a drill, to do all the work, in fact, by the dearest mode instead of the cheapest. The destroyers of property have got the law, therefore, into their own hands, as far as their triumph over machinery is concerned. And how do they proceed in their career? The farmer, we have imagined, takes all this quietly; he pays the new labourers who have got into his barns and his fields, ready for hand-work, with their flails and their dibbers; but he employs *just as many people as are absolutely necessary*, and no more, for getting his corn ready for market, and for preparing, in a slovenly way, for the seed-time. In a month or two, the victorious destroyers find that not a single hand the more of them is really employed. And why not? There are no drainings going forward, the hedges and ditches are neglected, the dung-heap is not turned over, the chalk is not fetched from the pit; in fact, all those labours are neglected which belong to a state of agricultural industry which is brought to perfection. *The farmer has no funds to employ in such labours;* he is paying a great deal more than he paid before, because his labourers choose to do certain labours with rude tools, instead of perfect ones. If he is a humane man, but not a firm one, and yields to the notion that it is a good thing to do work in the most round-about way, and, consequently, at the greatest expense, he does not regret the loss of the drill-plough, which, as well as many other agricultural tools, lessens labour, lessens the quantity of seed, and increases the crop. He spends all the saving which the drill-plough has caused upon the clamorous labourers, who insist that their arms are better for his work than the instrument which they have broken. But still some work must be neglected, and thus the actual amount of money paid for labour is just the same.

We will imagine that this state of things continues till

the next spring. All this while the price of grain has been rising. Many farmers have ceased to employ capital at all upon the land. The neat inventions, which enabled them to make a living out of their business, being destroyed, they have abandoned the business altogether. Others, who have yielded to the uproar, go on as well as they can, neglecting a great many labours that unite to make a good crop, although they are more and more pressed upon by the demand for labour, in consequence of a deal of land going out of cultivation. They pay as long as they can; they pay much more for labour than they did before, some out of compassion, and some out of fear. But the prices are steadily rising, and, therefore, although there is more money paid to a greater number of labourers, and although some receive higher wages, in the common sense of the word, than they did before they broke the machines, they are infinitely worse off, for the rate of wages is really lower. A day's work will then no longer purchase as much bread as before. It requires the work of a day and a half to procure the same quantity. Wages are, therefore, really lower, because a less crop is being produced at a greater cost, and the market-price is influenced accordingly.

The labourers now either begin to quarrel amongst themselves, and give up their unwise combinations; or they still combine, with the determination to obtain employment by the destruction of everything that appears to them to stand in the way of it,—everything beyond the teeth and nails of the workmen, as Joseph Foster expressed himself,—and on they go in the work of ruin. The horse, it may be probably found out, is as great an enemy as the drill-plough; so the horses are turned out to starve, or have their throats cut, the laws being still idle. This is, indeed, a great point gained, for, as a horse will do the field-work of six men, there must be six men employed, without doubt, instead of one horse: so would conclude these most mistaken violators of the laws by which society is held together. But how would the fact turn out? If the farmer still went on, in spite of all these losses and crosses, he might employ men in the place of horses, but not a single man more than the number that would work at the price of the keep of one horse. To do the work of each horse destroyed he would require six men ; but he would only have about a shilling a-day to divide between these six—the amount which the horse consumed.

In the meantime it would be perfectly evident, from all this convulsion amongst the labourers, from all this wanton and profitless ruin, that a great deal of the land would very quickly go out of cultivation altogether, if the laws were still idle. While the land was going out of cultivation, the stock of corn on hand would be much more quickly decreasing than in quiet times. The thrashing-machine is held to produce some saving of corn, as compared with the flail. The destruction of drills, also, would cause a larger quantity of seed to be sown than would be otherwise necessary. Thrashing-machines and drills, taken together in their savings, prevent therefore some thousand quarters of wheat alone from being wasted. Fifteen million quarters of wheat are annually consumed in Great Britain. If they save only a fiftieth of the wheat, as much, therefore, as would feed all the people of Great Britain for a week, or three hundred thousand of those people all the year, would be absolutely thrown away, trodden under foot, sent to the dunghill with the straw, wasted for ever, by one wrong act alone of the labourers ; and all this while we should be preparing to grow a great deal less for the next harvest. We should have a famine, if foreign countries could not supply us with what the labourers had destroyed. And how do we know that foreign labourers would be wiser than English ones, and abstain from such acts, in the knowledge that whatever raises the price of produce lowers the rate of wages ? Are foreign labourers better informed than English labourers ? If so, let us all take shame to ourselves, that the means of acquiring knowledge which this country affords have been neglected ; for upon sound knowledge must rest the safety and happiness of all.

About three or four hundred years ago, from the times of king Henry IV. to those of king Henry VI., and, indeed, long before these reigns, there were often, as we have already mentioned, grievous famines in this country, because the land was very wretchedly cultivated. Men, women, and children perished of actual hunger by thousands; and those who survived kept themselves alive by eating the bark of trees, acorns, and pig-nuts. There were no machines then; but the condition of the labourers was so bad, that they could not be kept to work upon the land without those very severe and tyrannical laws noticed in Chap. VII., which absolutely forbade them to leave the station in which they were born as labourers, for any hope of bettering their condition in the towns. There were not labourers enough to till the ground, for they worked without any skill, with weak ploughs and awkward hoes. They were just as badly off as the people of Portugal and Spain at our own day, who are miserably poor, *because* they have bad machines; or as the Chinese labourers, who have scarcely any machines, and are the poorest in the world. There was plenty of labour to be performed, but the tools were so bad, and the want of agricultural knowledge so universal, that the land was never half cultivated, and therefore all classes were poorly off. They had little corn to exchange for manufactures, and in consequence the labourer was badly clothed, badly lodged, and had a very indifferent share of the scanty crop which he raised. In the natural course of things, a good deal of land was laid down to grass; this was superseding labour to a great extent, and much clamour was raised about this plan, and probably a good deal of real distress was produced. But mark the consequence. Although the money wages of labour were lowered, because there were more labourers in the market, the real amount of wages was higher, for better food was created by pasturage at a cheap rate. The labourer then got meat who had never tasted it before; and when the use of animal food became general, there were cattle and corn enough to be exchanged for manufactured goods, and the labourer got a coat and a pair of shoes, who had formerly gone half naked.

A very accurate French writer, M. Dupin, in a book, in which he enters into many comparisons between the condition of the people of England and that of the people of France, says, that two-thirds of the French people are at this day wholly deprived of the nourishment of animal food, and that they live wholly on chestnuts, or maize, or potatoes. He accounts for this by stating that in France only $7\frac{1}{2}$ parts in 100 of the soil are cultivated in meadows, while in England one-third of the whole country is in meadow, or 33 parts in 100. He says, therefore, that the inhabitants of England consume three times as much meat, milk, butter, and cheese, as the inhabitants of France; and that the people of England are consequently three times better furnished with good food. This Frenchman, who writes with an earnest desire to better the condition of his countrymen, exhorts them to improve the breed of cattle and to lay down more land to grass, that the people may be better nourished. If he thought that more labour, without increased production, would better the people, he would exhort them to break up the $7\frac{1}{2}$ parts in 100 of grass land, and go to work in raising more corn and more potatoes. He does no such thing. He knows that to lessen the price by increasing the quantity of animal food, or of any other comfort, is really to better the condition of the people, by really raising the wages of labour.

But to return to our triumphant machine-breakers, who utterly despise such considerations. As the year advanced, and the harvest approached, it would be discovered that not one-tenth of the land was sown: for although the ploughs were gone, because the horses were turned out to starve, and there was plenty of *labour* for those who chose to labour for its own sake, or at the price of a horse, this amazing employment for human hands, somehow or other, would not quite answer the purpose. It has been calculated that the power of horses, oxen, &c. employed in husbandry in Great Britain is ten times the amount of human power. If the human power

F

insisted upon doing all the work with the worst tools, the certainty is that not even one-tenth of the land could be cultivated. Where, then, would all this madness end? In the starvation of the labourers themselves. The people in the towns would probably use *their* machines, which are ships, and barges, and wagons, to bring them as much of the produce of foreign countries as they could get (and that would be little) in exchange for their manufactures; and the agricultural labourers, who had put themselves out of the sympathy and protection of society, if they were allowed to eat up all they had produced by such imperfect means, would be in an infinitely more wretched state than they could possibly be, if the demand for labour was many times less than it now is. They would be just in the condition of any other barbarous people, that were ignorant of the inventions that constitute the power of civilization. They would eat up the little corn which they raised themselves; and have nothing to give in exchange for clothes, and coals, and candles, and soap, and tea, and sugar, and all the many comforts which those who are even the worst off are not wholly deprived of.

All this may appear as an extreme statement; and certainly we believe that no such evils could happen: for if the laws were passive, the most ignorant of the labourers themselves would, if they proceeded to carry their own principle much farther than they have done, see in their very excesses the real character of the folly and wickedness to which it has led, and would lead them. Why should the labourers of Aylesbury not have destroyed the harrows as well as the drills? Why leave a machine which separates the clods of the earth, and break one which puts seed into it? Why deliberate about a horse-churn, when they are resolved against a winnowing-machine? The truth is, these poor men perceived, even in the midst of their excesses, the gross deception of the reasons which induced them to commit them. Their motive was a natural, and, if lawfully expressed, a proper impatience, under a condition which had certainly many hardships, and those hardships in great part produced by the want of profitable labour. But in imputing those hardships to machinery, they were at once embarrassed when they came to draw distinctions between one sort of machine and another. This embarrassment decidedly shews that there were fearful mistakes at the bottom of their furious hostility to machinery.

CHAPTER XIII.

It has been said by persons whose opinions are worthy attention, that spade-husbandry is, in some cases, better than plough-husbandry ;—that is, that the earth, under particular circumstances of soil and situation, may be more fitly prepared for the influences of the atmosphere, by digging, than by ploughing. It is not our business to enter into a consideration of this question. The growth of corn is a manufacture, in which man employs the chemical properties of the soil and of the air, in conjunction with his own labour, aided by certain tools or machines, for the production of a crop ; and that power, whether of chemistry or machinery,—whether of the salt, or the chalk, or the dung which he puts upon the earth, or the spade or the plough which he puts into it,—that power which does the work easiest is necessarily the best, *because it diminishes the cost of production.* If the plough does not do the work so well as the spade, it is a less perfect machine ; but the less perfect machine may be preferred to the more perfect, because, taking other conditions into consideration, it is a cheaper machine. If the spade, applied in a peculiar manner by the strength and judgment of the man using it, more completely turns up the soil, breaks the clods, and removes the weeds than the plough, which receives one uniform direction from man with the assistance of other animal power, then the spade is a more perfect machine in its combination with human labour than the plough is, worked with a lesser degree of the same combination. But still it may be a machine which cannot be used with advantage to the producer, and is therefore not desirable for the consumer. All such questions must be determined by the cost of production; and that cost in agriculture is made up of the rent of land, the profit of capital, and the wages of labour—or the portions of the produce belonging to the landlord, the farmer, and the labourer. Where rent is high, as in the immediate neighbourhood of large towns, it is important to have the labour performed as carefully as possible. It is then economy to turn the soil to the greatest account, and the land is cultivated as a garden. Where rent is low, it is important to have the labour performed with less care, because one acre cultivated by hand may cost more than two cultivated by the plough. It is then economy to save in the labour, and the land is cultivated as a field. In one case, the machine called a spade is used; in the other the machine called a plough is employed. The use of the one or the other belongs to practical agriculture, and is a question only of relative cost.

And this brings us to the great *principle* of all machinery. A tool of the simplest construction is a machine; a machine of the most curious construction is only a complicated tool. There are many cases in the arts, and there may be cases in agriculture, in which the human arm and hand, with or without a tool, may do work that no machine can so well perform. There are processes in polishing, and there is a process in copper-plate printing, in which no substance has been found to stand in the place of the human hand. And if, therefore, the man with a spade alone does a certain agricultural work more completely than a man guiding a plough, and a team of horses dragging it (which we do not affirm or deny), the only reason for this is, that the man with the spade is a better machine than the man with the plough and the horses. The most stupid man that ever existed is, beyond all comparison, a machine more cunningly made by the hands of his Creator, more perfect in all his several parts, and with all his parts more exquisitely adapted to the regulated movement of the whole body, less liable to accidents, and less injured by wear and

tear, than the most beautiful machine that ever was, or
ever will be, invented. There is no possibility of supplying in many cases a substitute for the simplest movements of man's body, by the most complicated movements
of the most ingenious machinery. And why so? Because the natural machinery by which a man even lifts
his hand to his head is at once so complex and so simple,
so apparently easy and yet so entirely dependent upon
the right adjustment of a great many contrary forces,
that no automaton, or machine imitating the actions of
man, could ever be made to effect this seemingly simple
motion, without shewing that the contrivance was imperfect,—that it was a mere imitation, and a very clumsy
one. What an easy thing it appears to be for a farming-
man to thrash his corn with a flail; and yet what an
expensive arrangement of wheels is necessary to produce
the same effects with a thrashing-machine! The truth
is, that the man's arm and the flail form a much more
curious machine than the other machine of wheels,
which does the same work; and the real question as re-
gards the value of the two machines is, which machine
in the greater degree lessens the cost of production?

We state this principle broadly, in our examination
into the value of machinery in diminishing the cost of
producing human food. A machine is not perfect, because it is made of wheels or cylinders, employs the
power of the screw or the lever, is driven by wind or
water or steam, but because it best assists the labour of
man, by calling into action some power which he does
not possess in himself. If we could imagine a man
entirely dispossessed of this power, we should see the
feeblest of animal beings. He has no tools which are a
part of himself, to build houses like the beaver, or cells
like the bee. He has not even learnt from nature to
build, instinctively, by certain and unchangeable rules.
His power is in his mind; and that mind teaches him to
subject all the physical world to his dominion, by avail-
ing himself of the forces which nature has spread around
him. To act upon material objects he arms his weakness
with tools and with machines. As we have before said,

tools and machines are in principle the same. When we strike a nail upon the head with a hammer, we avail ourselves of a power which we find in nature—the effect produced by the concussion of two bodies; when we employ a water-wheel to beat out a lump of iron with a much larger hammer, we still avail ourselves of the same power. There is no difference in the nature of the instruments, although we call the one a tool, and the other a machine. Neither the tool nor the machine have any force of themselves. In one case the force is in the arm, in the other in the weight of water which turns the wheel.

The chief distinction between man in a rude, and man in a civilized state of society is, that the one wastes his force, whether natural or acquired,—the other economizes, that is, saves it. The man in a rude state has very rude instruments; he therefore wastes his force: the man in a civilized state has very perfect ones; he therefore economizes it. Should we not laugh at the gardener who went to hoe his potatoes with a stick, having a short crook at the end? It would be a tool, we should say, fit only for children to use. Yet such a tool was doubtless employed by some very ancient nations; for there is an old medal of Syracuse which represents this very tool. The common hoe of the English gardener is a much more perfect tool, because it saves labour. Could we have any doubt of the madness of the man who would propose that all iron hoes should be abolished, to furnish more extensive employ to labourers who should be provided only with a crooked stick cut out of a hedge? The truth is, if the working men of England had no better tools than crooked sticks, they would be in a state of actual starvation. One of the chiefs of the people of New Zealand, who from their intercourse with Englishmen had learnt the value of tools, told Mr. Marsden, a missionary, that his wooden spades were all broken, and he had not an axe to make any more;—his canoes were all broken, and he had not a nail or a gimlet to mend them with;—his potato grounds were uncultivated, and he had not a hoe to break them up with;—and that *for want of cultivation* he and his people would have no-thing to eat. This shews the state of a people without tools.

But some would perhaps make a distinction, which we have endeavoured to show is a worthless one, between tools and machines. There are many who object to machinery, because, having grown up surrounded with the benefits it has conferred upon them, without understanding the source of these benefits, they are something like the child who sees nothing but evil in a rainy day. We have mentioned the people of New Zealand, who live exactly on the other side of the globe, and who, therefore, very rarely come to us; but when they do come they are acute enough to perceive the advantages which machinery has conferred upon us, and the great distance in point of comfort between their state and ours, principally for the reason that they have no machinery, while we have a great deal. One of these poor men burst into tears when he saw a rope-walk; because he perceived the immense superiority which the process of spinning ropes gave us over his own countrymen. Another of these people, and he was a very shrewd and intelligent person, carried back to his country a small hand-mill for grinding corn, which he prized as the greatest of all earthly possessions.

And well might he prize it! He had no machine for converting corn into meal, but two stones, such as were used in the remote parts of the highlands of Scotland some years ago. And to beat the grain into meal by these two stones (a machine, remember, however imperfect) would occupy the labour of one-fourth of his family, to procure subsistence for the other three-fourths. The ancient Greeks, three thousand years ago, had improved upon the machinery of the hand-stones, for they had hand-mills. But Homer, the old Greek poet, describes the unhappy condition of the slave who was always employed in using this mill. The groans of the slave were unheeded by those who consumed the produce of his labour; and such was the necessity for meal, that the women were compelled to turn these mills when there were not slaves enough taken in war to perform this irksome office. There was plenty of labour then to be performed, even with the machinery of the hand-mill; but the slaves and the women did not consider that labour was a good in itself, and therefore they bitterly groaned under it. By and bye, the understandings of men found out that water and wind would do the same work that the slaves and the women had done; and that a large quantity of labour was at liberty to be employed for other purposes. Some perhaps think that society was in a worse state in consequence. We answer, labour is worth nothing without results. Its value is only to be measured by what it produces. If in a country where hand-mills could be had, the people were to go on beating grain between two stones, all would pronounce them fools, because they could obtain an equal quantity of meal with a much less expenditure of labour. Some have a general prejudice against that sort of machinery which does its work with very little human assistance; it is not quite so certain, therefore, that they would agree that a people would be equal fools to use the hand-mill when they could employ the wind-mill or the water-mill. But we believe they would think, that if flour could drop from heaven, or be had like water by whoever chose to seek it, it would be the height of folly to have stones, or hand-mills, or water-mills, or wind-mills, or any machine whatever for manufacturing flour. Does any one ever think of *manufacturing* water? The cost of water is only the cost of the labour which brings it to the place in which it is consumed. Yet this admission overturns all objections against machinery. *We admit that it is desirable to obtain a thing with no labour at all; can we therefore doubt that it is desirable to obtain it with the least possible labour?* The only difference between no labour and a little labour, is the difference of the cost of production. And the only difference between little labour and much labour is precisely the same. In procuring any thing that administers to his necessities, man makes an exchange of his labour for the thing produced, and the less he gives of his labour the better of course is his bargain.

To return to the hand-mill and the water-mill. An ordinary water-mill for grinding corn will grind about thirty-six sacks a day. To do the same work with a hand-mill would require 150 men. At two shillings a day the wages of these men would amount to 15*l*., which, reckoning six working days, is 90*l*. a week, or 4680*l*. a year. The rent and taxes of a mill would be about 150*l*. a year, or ten shillings a working day. The cost of machinery would be certainly more for the hand-mills than the water-mill, therefore we will not take the cost of machinery into the calculation. To produce, therefore, thirty-six sacks of flour by hand we should pay 15*l*. ; by the water-mill we should pay ten shillings : that is, we should pay thirty times as much by the one process as by the other. The actual saving is something about two thirds of the price of the flour in the market; that is, the consumer, if the corn were ground by hand, would pay two thirds more than he pays now for flour ground at a mill.

But if the system of grinding corn by hand were a very recent system of society, and the introduction of so great a benefit as the water-mill had all at once displaced the hand-grinders, as the spinning machinery displaced the spinning-wheel, what must become, it is said, of the one hundred and fifty men who earned the 15*l*. a-day, of which sum the consumer has now got 14*l*. 10*s*. in his pocket ? They must go to other work. And what is to set them to that work ? The same 14*l*. 10*s*. ; which, being saved in the price of flour, gives the poor man, as well as the rich man, more animal food and fuel; a greater quantity of clothes, and of a better quality ; better furniture, and more of it ; domestic utensils ; and books. To produce these things there must be more labourers employed than before. The quantity of labour is, therefore, not diminished, while its productiveness is much increased. It is as if every man among us had become suddenly much stronger and more industrious. The machines labour for us, and are yet satisfied without either food or clothing. They increase all our comforts and they consume none themselves. The hand-mills are not grinding, it is true : but the ships are sailing that bring us foreign produce ; the looms are moving that give us more clothes ; the potter, and glass-maker, and joiner, are each employed to add to our household goods ; we are each of us elevated in the scale of society : and all these things happen because machinery has diminished the cost of production.

CHAPTER XIV.

ALL labourers in agriculture know full well the value of a tool; but some hate machinery. This is inconsistent. Unless the labourer made a plough (if he will consent even to a plough) out of two pieces of stick, and carried it upon his shoulder to the field, as the toil-worn and poor people of India do, he must have some *iron* about it. He cannot get iron without machinery. He hates machinery, and therefore he will have nothing to do with a plough! Will he have his hoe, then? He is not quite sure. Will he give up his knife? No; he must keep his knife. He has got every thing to do for himself, and his knife is his tool of all-work.

Well, how does he get this same knife? People that have no machinery sharpen a stone, or bit of shell, or bone, and cut or saw with it in the best way they can; and after they have become very clever, they fasten it to a wooden handle with a cord of bark. An Englishman examines two or three dozens of knives, selects which he thinks the best, and pays a shilling for it, the seller thanking him for his custom. The man who has nothing but the bone or the shell would gladly toil a month, for that which does not cost an English labourer half a day's wages.

And how does the Englishman obtain his knife upon such easy terms? From the very same cause that he obtains all his other accommodations cheaper, in comparison with the ordinary wages of labour, than the inhabitant of most other countries,—that is, from the use of machinery, either in the making of the thing itself, or in procuring that without which it could not be made. We must always remember that if we could not get the materials without machinery, it would be as impossible for us to get what is made of those materials as if a machine delivered it to us ready for use.

Keeping this in mind, let us see how a knife could be obtained by a man who had nothing to depend upon but his hands.

Ready-made, without the labour of some other man, a knife does not exist; but the iron, of which the knife is made, is to be had. Very little iron has ever been found in a native state, or fit for the blacksmith. The little that has been found in that state has been found only very lately; and if human art had not been able to procure any in addition to that, gold would have been cheap as compared with iron.

Iron is, no doubt, very abundant in nature; but it is always mixed with some other substance that not only renders it unfit for use, but hides its qualities. It is found in the state of what is called *iron stone*, or *iron ore*. Sometimes it is mixed with clay, at other times with lime or with the earth of flint; and there are also cases in which it is mixed with sulphur. In short, in the state in which iron is frequently met with, it is a much more likely substance to be chosen for paving a road, or building a wall, than for making a knife.

But suppose that the man knows the particular ore or stone that contains the iron, how is he to get it out? Mere force will not do, for the iron and the clay, or other substance, are so nicely mixed, that though the ore were ground to the finest powder, the grinder is no nearer the iron than when he had a lump of a ton weight.

A man who has a block of wood has a wooden bowl in the heart of it; and he can get it out too by labour. The knife will do it for him in time; and if he take it to the turner, the turner with his machinery, his lathe, and his gouge, will work it out for him in half an hour. The man who has a lump of iron ore has just as certainly a knife in the heart of it; but no mere labour can work it out. Shape it as you may, it is not a knife, or steel, or even iron,—it is iron ore; and dress it as you will, it would not cut better than a brick-bat,—certainly not so well as the shell or bone of the savage.

There must be knowledge before any thing can be done in this case. We must know what is mixed with the iron, and how to separate it. We cannot do it by mere labour, as we can chip away the wood and get out the bowl; and therefore we have recourse to fire.

In the ordinary mode of using it, fire would make matters worse. If we put the material into the fire as a stone, we should probably receive it back as slag or dross. We must, therefore, prepare our fuel. Our fire must be hot, very hot; but if our fuel be wood we must burn it into charcoal, or if it be coal into coke.

The charcoal, or coke, answers for one purpose; but we have still the clay or other earth mixed with our iron, and how are we to get rid of that? Pure clay, or pure lime, or pure earth of flint, remains stubborn in our hottest fires; but when they are mixed in a proper proportion, the one melts the other.

So charcoal or coke, and iron stone or iron ore, and limestone, are put into a furnace; the charcoal or coke is lighted at the bottom, and wind is blown into the furnace, at the bottom also. If that wind is not sent in by machinery, and very powerful machinery too, the effect will be little, and the work of the man great; but still it can be done. Within these few years the quality of the wind sent in, a hot-blast or a cold-blast, has made a great difference in the facility of producing iron. The labour is less, the iron is cheaper, because the wind employed in its production has been rendered more efficient.

In this furnace the lime and clay, or earth of flint, unite, and form a sort of glass, which floats upon the surface. At the same time the carbon, or pure charcoal, of the fuel, with the assistance of the limestone, mixes with the stone, or ore, and melts the iron, which, being heavier than the other matters, runs down to the bottom of the furnace, and remains there till the workman lets it out by a hole made at the bottom of the furnace for that purpose, and plugged with sand. When the workman knows there is enough melted, or when the appointed time arrives, he displaces the plug of sand with an iron rod, and the melted iron runs out like water, and is conveyed into furrows made in sand, where it cools, and the pieces formed in the principal furrows are called " sows," and those in the furrows branching from them, " pigs."

But great as is the advantage of this first step of the iron-making, the iron is not yet fit for a knife. It is cast-iron. It cannot be worked by the hammer, or sharpened to a cutting edge; and so it must be made into malleable iron,—into a kind of iron which, instead of melting in the fire, will soften, and admit of being hammered into shape, or united by the process of welding.

The methods by which this is accomplished vary; but they in general consist in keeping the iron melted in a furnace, and stirring it with an iron rake, till the blast of air in the furnace burns the greater part of the carbon out of it. By this means it becomes tough; and, without cooling, is taken from the furnace and repeatedly beaten by large hammers, or squeezed through large rollers, until it becomes the bar-iron of which so much use is made in every art of life.

Bringing it into this state requires great force; and the unaided strength of all the men in Britain could not make all the iron which is at present made, though they did nothing else. Machinery is therefore resorted to; and water-wheels, steam-engines, and all sorts of powers are set to work in moving hammers, turning rollers, and drawing rods and wires through holes, till every workman can have the particular form which he wants. If it were not for the machinery that is employed in the manufacture, no man could obtain a spade for less than the price of a year's labour; the yokes of a horse would cost more than the horse himself; and the farmer would have to return to wooden plough-shares, and hoes made of sticks with crooked ends. There would be *labour* enough then, as we have already shown :—but the people could not live upon the labour only; they must have *profitable* labour.

After all this, the iron is not yet fit for a knife, at least for such a knife as an Englishman may buy for a shilling. Many nations would, however, be thankful for a little bit of it, and nations too in whose countries there is no want of iron ore. But they have no knowledge of the method of making iron, and have no furnaces or machinery. When our ships sail among the people of the eastern islands, those people do not ask for gold. "Iron, iron!" is the call; and he who can exchange his best commodity for a rusty nail or a bit of iron hoop is a fortunate individual.

We are not satisfied with that in the best form, which is a treasure to those people in the worst. We must have a knife, not of iron, but of *steel*,—a substance that will bear a keen edge without either breaking or bending. In order to get that, we must again change the nature of our material.

How is that to be done? The oftener that iron is heated and hammered, it becomes the softer and more ductile; and as the heating and hammering forced the carbon out of it, if we give it the carbon back again, we shall harden it; but it happens that we also give it other properties, by restoring its carbon, when the iron has once been in a ductile state.

For this purpose, bars or pieces of iron are buried in powdered charcoal, covered up in a vessel, and kept at a red heat for a greater or less number of hours, according to the object desired. There are niceties in the process, which it is not necessary to explain, that produce the peculiar quality of steel, as distinguished from cast-iron. If the operation of heating the iron in charcoal is continued too long, or the heat is too great, the iron becomes cast steel, and cannot be welded; but if it is not melted in the operation, it can be worked with the hammer in the same manner as iron.

In each case, however, it has acquired the property upon which the keenness of the knife depends; and the chief difference between the cast steel and the steel that can bear to be hammered is, that cast steel takes a keener edge, but is more easily broken.

The property which it has acquired is that of bearing to be tempered. If it be made very hot, and plunged into cold water, and kept there till it is quite cooled, it is so hard that it will cut iron, but it is brittle. In this state, the workman brightens the surface, and lays the steel upon a piece of hot iron, and holds it to the fire till it becomes of a colour which he knows from experience is a test of the proper state of the process. Then he plunges it again into water, and it has the degree of hardness that he wants.

The grinding a knife, and the polishing it, even when it has acquired the requisite properties of steel, if they were not done by machinery, would cost more than the whole price of a knife upon which machinery is used. A travelling knife-grinder, with his treadle and wheels, has a machine, but not a very perfect one. The Sheffield knife-maker grinds the knife at first upon wheels of immense size, turned by water or steam, and moving so quickly that they appear to stand still—the eye cannot follow the motion. With these aids the original grinding and polishing cost scarcely any thing; while the travelling knife-grinder charges two-pence for the labour of himself and his wheel, in just sharpening it.

As iron is with us almost as plentiful as stone, we do not think much about it. But there is a great deal to be done, much thinking and inventing, before so simple a thing as a shilling knife could be procured : and without the thinking and the inventing all the strength of all the men that ever lived never could procure it ; and without the machinery to lighten the labour, no ingenuity could furnish it at a thousand times the expense.

And why then can an ordinary workman procure it for the price of a few hours' labour ? The causes are easily seen. Every part of the labour that can be done by machinery is so done. One turn of a wheel, one stroke of a steam-engine, one pinch of a pair of rollers, or one blow of a die, will do more in a second than a man could do in a month. Another advantage in the manufacture is that the labour is divided. Each man has but one thing to do, and in course of time he comes to do twenty times as much as if he were constantly shifting from one thing to another. The value of the work that a man does is not to be measured in all cases by the time and trouble that it costs him individually, but by the market value of what he produces; which value is determined, as far as labour is concerned, by the price paid for doing it in the best and most expeditious mode. He that fells a sapling by one stroke of his axe, does just as much work as he that should take to rip it asunder with his nails, or gnaw it with his teeth, according to Joseph Foster's expression.

And does not all this machinery, it may still be said, deprive many workmen of employment? No. By these means the iron trade gives bread to hundreds, where otherwise it would not have given bread to one. There are more hands employed at the iron-works than there would have been if there had been no machinery; because without machinery men could not produce iron cheap enough to be generally used.

Then the number of founders and smiths! The pans and kettles, the pipes, the grates, the thousands of iron things that we see, all employ somebody; and every body is benefited by the use of them. We make bridges of iron, roads of iron, waggons of iron, boats of iron, steam-engines of iron; and, in fact, so many things of iron, that if we were deprived of it, our country would not be worth inhabiting. As many iron pipes have been laid down for supplying water and light to the inhabitants of London alone, within the last twenty years, as, without machinery, would have kept all the world busy for a century, or rather could not have been procured at all.

Then there are the tools of every trade. All the working parts of them are made of iron, and very many of them are made of iron altogether. How would the hedger proceed without his bill, or the ditcher without his mattock? The first could not do with wood at all, and the last would make but sorry work. Then come the spade, the hoe, the scythe, and the sickle. How could they be had without iron ? or what would be the condition of the labourer, if iron and iron tools were sold at the price of making them without machinery ? His tools now cost him only the produce of a few days of his own labour ;—in many cases they are supplied to him. If they were made by hand, they would cost him the labour of many weeks to purchase them, if they could be purchased at all. His tools are, however, a first necessary of life to the labourer ;—he cannot earn his bread without them. If he bought the tools at the dear rate, he would probably spend half his earnings in buying them ; and then he must be paid double.

But instead of there being any thing to pay double wages to the labourer, every body would be in the same necessity as himself ; and the necessity would be that he and they would either pay double for all that was bought, or work for half price. The axe, the saw, the plane, the gimlet, and the nails, would consume all the earnings of the carpenter ; the needles, the shears, and the goose, would be a burden to the tailor ; the farmer would be obliged to pay in the additional price of iron what he now pays to his labourers, and labour itself would be at an end.

There was a time when iron was made in this country with very little machinery. Iron was manufactured here in the time of the Romans ; but it was made with great manual labour, and was consequently very dear. Hutton, in his History of Birmingham, tells us that there is a large heap of cinders near that town which have been produced by an ancient iron furnace ; and that from the quantity of cinders, as compared with the mechanical powers possessed by our forefathers, the furnace must have been constantly at work from the time of Julius Cæsar. A furnace with a steam blast would produce as large a heap in a few years. It is always a difficulty at the iron-works to get rid of their cinders.

The machinery that is now employed in the iron trade, not only enables the people to be supplied cheaply with all sorts of articles of iron, but it enables a great number of people to find employment, not in the iron trade only, but in all other trades, who otherwise could not have been employed; and it enables everybody to do more work with the same exertion by giving them better tools; while it makes all more comfortable by furnishing them with more commodious domestic utensils.

There are thousands of families on the face of the earth, that would be glad to exchange all they have for a tin kettle, or an iron pot, which can be bought any where in the three kingdoms for eighteen pence. And could the poor man in this country but once see how even the rich man in some other places must toil day after day before he can scrape or grind a stone so as to be able to boil a little water in it, or make it serve for a lamp, he would account himself a poor man no more. An English gipsy carries about with him more of the conveniences of life than are enjoyed by the chiefs or rulers in countries which naturally have much finer climates than that of England. But they have no machinery, and therefore they are wretched.

Nearly all the people now engaged in iron-works are supported by the improvements that have been made in it *by machinery* since 1788. Yes, wholly by the machinery; for before then the quantity made by the charcoal of wood had fallen off one-fourth in forty-five years. The wood for charcoal was becoming exhausted, and nothing but the powerful blast of a machine will make iron with coke. Without the aid of machinery the trade would have become extinct. The iron and the coal employed in making it would have remained useless in the mines.

CHAPTER XV.

In every part of the operation of making iron;—in smelting the iron out of the ore; in moulding cast iron into those articles for which it is best adapted; in working malleable iron and steel, and in applying them to use after they are made; nothing can be done without fire, and the fuel that is used in almost every stage of the business is coal. The coal trade and the iron trade are thus so intimately connected, so very much dependent upon each other, that neither of them could be carried on to any extent without the other. The coal mines supply fuel, and the iron-works give mining tools, pumps, railroads, wheels, and steam-engines, in return. A little coal might be got without the iron engines, and a little iron might be made without coals, by the charcoal of wood. But the quantity of both would be trifling in comparison. Thus it is, that to make one million five hundred thousand tons of iron, requires above three times the quantity, or five million tons of coal. This large quantity is only for the production of pig-metal. In other castings, and in the various subsequent stages of iron manufacture, many more tons of coal are employed.

Through the general consumption of wood instead of coal, a fire for domestic use in France is a great deal dearer than a fire in England; because, although the coal-pits are not to be found at every man's door, nor within many miles of the doors of some men, machinery at the pits, and ships and barges, which are also machinery, enable most men to enjoy the blessings of a coal fire at a much cheaper rate than a fire of wood, which is not limited in its growth to any particular district. Without the machinery to bring coals to his door, one man out of fifty of the present population of England could not have had the power of warming himself in winter; any more than without the machines and implements of farming he could obtain food, or without those of the arts he could procure clothing. The sufferings produced by a want of fuel cannot be estimated by those who have abundance. In Normandy, very recently, such was the scarcity of wood, that persons engaged in various works of hand, as lace-making by the pillow, absolutely sat up through the winter nights in the barns of the farmers, where cattle were littered down, that they might be kept warm by the animal heat around them. They slept in the day, and were warmed by being in the same out-house with cows and horses at night;—and thus they worked under every disadvantage, because fuel was scarce and very dear.

Well, suppose that our machinery (not mere tools, but more complicated instruments) were laid aside, and all the works and contrivances that have been made by means of it put an end to, and the people left to find coals with shovels and pickaxes, and all the tools used in common digging. How would they go about this task, or what would the coals cost as compared with the present price?

At present a cottager in the south of England, where there is no coal in the earth, may have a bushel of good coals delivered at the door of his cottage for eighteen pence. If he had even the means of transporting himself and his family to the coal district, he could not, without machinery, get a bushel of coals at the price of a year's work. Let us see how a resolute man would proceed in such an undertaking.

The machinery, we will say, is gone. The mines are filled up, which the greater part of them would be, with water, if the machinery were to stop a single week. Let us suppose that the adventurous labourer knows exactly the spot where the coal is to be found. This knowledge, in a country that has never been searched for coals before, is no easy matter even to those who understand the subject best; but we shall suppose that he gets over that difficulty too, for after it there is plenty of difficulty before him.

Well, he comes to the exact spot that he seeks, and places himself right over the seam of coal. That seam is only a hundred fathoms below the surface, which depth he will, of course, reach in good time. To work he goes; pares off the green sod with his shovel, loosens the earth with his pickaxe, and, in the course of a week, is twenty feet down into the loose earth and gravel, and clears the rock at the bottom. He rests during the Sunday, and comes refreshed to his work on Monday morning; when behold there are twelve feet of water in his pit.

Somebody now lends him a bucket and rope (a machine, mind), and he bales away, till, as night closes, he has lowered the water three feet. Next morning it is up a foot and a half: but no matter; he has done something, and next day he redoubles his efforts, and brings the water down to only four feet. That is encouraging; but, from the depth, he now works his bucket with more difficulty, and it is again a week before his pit is dry. The weather changes; the rain comes down heavily; the surface on which it falls is spongy; the rock which he has reached is water-tight; and in twelve hours his pit is filled to the brim. It is in vain to go on.

The sinking of a pit, even to a less depth than a hundred fathoms, sometimes demands, notwithstanding all the improvements by machinery, a sum of not less than a hundred pounds a fathom, or ten thousand pounds for the whole pit; and therefore, supposing it possible for a single man to do it at the rate of eighteen pence a day, the time which he would require would be between four hundred and five hundred years.

Whence comes it that the labour of between four hundred and five hundred years is reduced to a single day, and that which, independently of the carriage, would have cost ten thousand pounds, is got for eighteen pence?

It is because man joins with man, and machinery is employed to do the drudgery. Nations that have no machinery have no coal fires, and are ignorant that there is
hidden under the earth a substance which contributes
more, perhaps, to the health and comfort of the inhabitants of Britain than any other commodity which they
enjoy.

Yes, that the coals are procured for so mere a trifle in
comparison—that they are procured at all—is owing to
every part of the labour that can be so performed being
performed by machinery; and that the machinery exists,
and this vast addition is made to every man's comfort, is
owing to there being men in the country with capital
enough to bear the expense, and with spirit enough to
risk their property in operations of so much cost and
magnitude.

Some of the particulars of the mode in which coals are
obtained deserve to be mentioned, for the purpose of
shewing with what ease machinery can do that which to
human strength, however skilfully applied, if unassisted
by machinery, would be utterly impossible.

Suppose the men digging in a narrow pit, such as that
which is sunk for the purpose of "winning" coals, how
would they, after they had reached a certain depth, raise
the earth and rubbish to a surface, and go down or come
up?   Without machinery there are really no means, but
by putting down a ladder, or cutting steps in the side of
the pit.   With such aids, let us consider how much rubbish a man could bring up in a day.   Coulomb, a French
mathematician, found that a man without any load could
climb up steps to a height of about eighty-two fathoms,
eighteen times within the hours of ordinary work. There
would be therefore, at most, only eighteen daily ascents
from the pit, even if the man brought nothing with him.
If he carry a load, his power of ascent is of course much
diminished.

But art performs what mere bodily strength cannot
accomplish.   When the pit becomes too deep for throw-
ing out the earth with a shovel, a windlass is put across

it with a rope and buckets, so that while the full one is drawn up the empty one is let down. One man works the windlass, and another digs.

When the man who digs comes to stone, he lays aside the pickaxe and the shovel, takes the iron "jumper," drills a hole in the rock, puts in a shot of gunpowder, lays a few faggots over for safety, lights the train, and is drawn some yards up by the bucket. Thunder goes the shot, the faggots prevent all danger from the fragments, and as much stone is loosened in two minutes as the man would have loosened with his pickaxe in a week. This is a chemical aid; an aid which is as powerful in diminishing cost of production as the aid of machinery. There is therefore now nothing to do but to fill the buckets, and so another man is sent to the windlass.

When the pit gets still deeper, the windlass is laid aside, and the horse gin is mounted. According to Smeaton, the power of a horse is equal to five men at the same kind of labour. Other writers estimate the power of a horse at more than six men. One of the three men can attend to the horse and empty the buckets, and the two are at liberty to dig and fill below; so that with the gin and horse, the labour of many men is saved.

The number of men that can work at a windlass, or the number of horses that can be yoked to a gin, is limited. But the power of the steam-engine is limited only by the strength of the materials of which it is formed. The power of a hundred horses, or of five hundred men, may be very easily made by the steam-engine to act constantly, and on a single point; and thus there is scarcely any thing in the way of mere force which the engine cannot be made to do. We have seen a pit in Staffordshire, which hardly gave coal enough to maintain a cottager and his family, for he worked the pit with imperfect machinery; with a half-starved ass applied to a windlass. A mile off was a steam-engine of 200-horse power, raising tons of coals and pumping out rivers of water with a force equal to at least a thousand men. This vast force acted upon a point; and therefore no advantage was gained over the machine by the opposing force of water,

or the weight of the material to be raised.   Before the
steam-engine was invented, the produce of the coal-
mines barely paid the expense of working and keeping
them dry ; and had it not been for the steam-engines and
other machinery, the supply would long before now have
dwindled into a very small quantity, and the price would
have become ten or twenty times its present amount.

Engines or machines, of some kind or other, not only
keep the pits dry and raise the coals to the surface, but
convey them to the ship upon railroads ; and the ship,
itself a machine, carries them round all parts of the
coast ; while barges and boats (still machines) convey
them along the rivers and canals.   By these means, a
ton is carried from the pit to the cottage of the labourer,
for less expense than, in the distant places, he himself
could fetch a pound, even although he were to get it for
nothing at the pit.   We shall see, in the next chapter,
how all machines which facilitate the general communication between man and man, whether from town to
town, from district to district, from county to county, or
from nation to nation, advance the general welfare of
mankind.

What are the consequences of machinery employed in
raising coals ?   That by machinery many millions of tons
of one of the very first necessaries of life are obtained,
which without machinery could not be obtained at all in
the thousandth part of the quantity ; and which, consequently, would be a thousand times the price—would, in
fact, be precious stones, instead of common fuel.

CHAPTER XVI.

THE chief power which produces coal and iron cheap at the pit is that of machinery. It is the same power which distributes these bulky articles through the country, and equalizes the cost in a considerable degree to the man who lives in London and the man who lives in Durham or Staffordshire. The difference in cost is the price of transport; and machinery, applied in various improved ways, is every year lessening the cost of conveyance, and thus equalizing prices throughout the British Islands. The same applications of mechanical power enable a man to move from one place to another with equal ease, cheapness, and rapidity. Quick travelling has become cheaper than slow travelling. The time saved remains for profitable labour.

About a hundred and sixty years ago, when the first turnpike-road was formed in England, a mob broke the toll-gates, because they thought an unjust tax was being put upon them. They did not perceive that this small tax for the use of a road would confer upon them innumerable comforts, and double and treble the means of employment.

If there were no road, and no bridge, a man would take six months in finding his way from London to Edinburgh, if indeed he found it at all. He would have to keep the line of the hills, in order that he might come upon the rivers at particular spots, where he would be able to jump over them with ease, or wade through them without danger.

When a man has gone up the bank of a river for

twelve miles in one direction, in order to be able to cross it, he may find that before he proceeds one mile in the line of his journey, he has to go along the bank of another river for twelve miles in the opposite direction ; and the courses of the rivers may be so crooked that he is really farther from his journey's end at night than he was in the morning.

He may come to the side of a lake, and not know the end at which the river, too broad and deep for him to cross, runs out ; and he may go twenty miles the wrong way, and thus lose forty.

Difficulties such as these are felt by every traveller in an uncivilized country. In reading books of travels, in Africa for instance, we sometimes wonder how it is that the adventurer proceeds a very few miles each day. We forget that he has no roads.

Two hundred years ago—even one hundred years ago —in some places fifty years ago—the roads of England were wholly unfit for general traffic and the conveyance of heavy goods. Pack-horses mostly carried on the communication in the manufacturing districts. The roads were as unfit for moving commodities of bulk, such as coal, wool, and corn, as the sandy roads of Poland are at the present day. Bad roads in Poland double the original price of wheat by the cost of conveyance a very few miles. Bad corn-laws in England prevent the natural course of commercial exchange, which would very soon mend the Polish roads from the corn-field to the sea-port. The great principle of exchange between one part of this island and another part, which has ceased to be an affair of restrictions and jealousies, has covered this island with good roads, with canals, and finally with railways. The railway and the steam-carriage have carried the principle of diminishing the price of convey-ance, and therefore of commodities, by machinery, to an extent which makes all other illustrations almost unneces-sary. A road with a waggon moving on it is a mecha-nical combination ; a canal, with its locks and towing-paths, and boats gliding along almost without effort, is a higher mechanical combination ; a railway, with its locomotive engine, and carriage after carriage dragged along at the rate of thirty miles an hour, is the highest of such mechanical combinations. The force applied upon a level turnpike-road, which is required to move 1800 lbs., if applied to drag a canal boat, will move 55,500 lbs., both at the rate of 2½ miles per hour. But we want economy in time as well as economy in the application of motive power. It has been attempted to apply speed to canal travelling. Up to 4 miles an hour the canal can convey an equal weight more economically than a railroad ; but after a certain velocity is exceeded, that is 13½ miles an hour, the horse on the turnpike-road can drag as much as the canal team. Then comes in the great advantage of the railroad. The same force that is required to draw 1900 lbs. upon a canal, at a rate above 13½ miles an hour, will draw 14,400 lbs. upon a railway, at the rate of 13½ miles an hour. Who can doubt that the cost of consumption is diminished by machinery, when the producers and consumers are thus brought together, not only at the least cost of transit, but at the least expenditure of time ?

If we add to the road, the canal, and the railway, the steam-boat traffic of our own coasts, we cannot hesitate to believe that the whole territory of Great Britain and Ireland is more compact, more closely united, more accessible, than was a single county two centuries ago. It may be said, without exaggeration, that it would now be impossible for a traveller in England to set himself down in any situation where the post from London would not reach him in eighteen hours. When the first edition of ' The Results of Machinery ' was published in 1831, we said that the post from London would reach any part of England in three days ; and that " fifty years before, such a quickness of communication would have been considered beyond the compass of human means." In fourteen years we have so diminished the practical amount of distance between one part of Great Britain and another, that the post from London to Aberdeen is only thirty-six hours. In a few years it will be even less. Railways are producing these wonderful changes ; and in connexion with railways and improved roads and steam-ships, the mental labourers have been at work with improved organization to make the condition of all other labourers more advantageous.

Roads, canals, steam-ships, railways, are each and all machines for diminishing the cost of transport, whether of commodities or of human beings. They create labour, they lower and equalize prices. About twenty years ago a new road was made, at the expense of government, through a barren country, which presented an impassable barrier to communication between Limerick, Cork, and Kerry. We will take one example of the instantaneous benefit of this road-making, as described by a witness before Parliament :—" A hatter, at Castle-island, had a small field through which the new road passed : this part next the town was not opened until 1826. In making arrangements with him for his damages, he said that he ought to make me (the engineer) a present of all the land he had, for that the second year I was at the roads he sold more hats to the people of the mountains alone, than he did for seven years before to the high and low lands together."

The hatter of Castle-island got comfort and prosperity by the roads, because the man who had to sell and the man who had to buy were brought closer to each other by means of the roads. When there were no roads, the hatter kept his goods upon the shelf, and the labourer in the mountains went without a hat. When the labourer and the hatter were brought together by the roads, the hatter soon sold off his stock, and the manufacturer of hats went to work to produce him a new stock ; while the labourer, who found the advantage of having a hat, also went to work to earn more money, that he might pay for another when he should require it. It became a fashion to wear hats, and of course a fashion to work hard, and to save time, to be able to pay for them. Thus the road created industry on both sides, on the side of the producer of hats and that of the consumer.

What the new Irish road did for the hatter of Castle-island, the railroads of England and Scotland have done, and are doing, for our millions of producers and consumers. But it may be held by some that railways, as far as passenger communication goes, are inventions for the benefit of the rich and the pleasure-seeking. Parliament thought otherwise when it enacted, in 1844, that upon every railway there should be a train once a day provided for third-class passengers, in carriages with seats, and protected from the weather, which should take up and set down passengers at every station, and the fare not to exceed one penny per mile. If all railway proprietors had understood their own interests, none would have waited for a legislative enactment to carry third-class passengers at a penny a mile. But before this act of Parliament was passed, the penny-a-mile passengers formed an important class of travellers. From the 1st of July, 1842, to the 30th of June, 1843, sixty-six railways then in operation in the United Kingdom received from passengers and for the conveyance of goods the enormous sum of four million five hundred and thirty-five thousand pounds. Of this sum three million one hundred and ten thousand pounds was received from passengers, amounting to more than twenty-three million persons. Of these, six million five hundred thousand were third-class passengers, who paid four hundred and eleven thousand pounds, being upon an average fifteen pence for each. It is evident that the third-class passengers went short distances, some less than fifteen miles, some more. Can any one doubt that the *free interchange of labour* is promoted in an unexampled degree by such railway communication?

When William Hutton, in the middle of last century, started from Nottingham (where he earned a scanty living as a bookbinder) and walked to London and back for the purpose of buying tools, he was nine days from home, six of which were spent in going and returning. He travelled on foot, dreading robbers, and still more dreading the cost of food and lodging at public-houses. His whole expenses during this toilsome expedition were only ten shillings and eight pence; but he contented himself with the barest necessaries, keeping the money for his tools sewed up in his shirt-collar. If William Hutton had lived in these days, he would, upon sheer principles of economy, have gone to London and back by the Nottingham train in two days, at a cost of twenty shillings for his transit. The twenty shillings would have been sacrificed for his conveyance, but he would have had a week's labour free to go to work with his new tools; he need not have sewed his money in his shirt-collar for fear of thieves; and his shoes would not have been worn out and his feet blistered in his toilsome march of two hundred and fifty miles.

And there are some men who say that this wonderful communication, the greatest triumph of modern skill, is not a blessing;—for the machinery has put somebody out of employ. Baron Humboldt, a traveller in South America, tells us, that upon a road being made over a part of the great chain of mountains called the Andes, the government was petitioned against the road, by a body of men who for centuries had gained a living by carrying travellers in baskets strapped upon their backs over the fearful rocks, which only these guides could cross. Which was the better course—to make the road, and create the thousand employments belonging to freedom of intercourse, for these very carriers of travellers, and for all other men; or to leave the mountains without a road, that the poor guides might gain a premium for risking their lives in an unnecessary peril?

CHAPTER XVII.

THE people who live in small villages, or in scattered habitations in the country, have certainly not so many *direct* benefits from machinery as the inhabitants of towns. They have the articles at a cheap rate, which machines produce, but there are not so many machines at work for them as for dense populations. From want of knowledge they may be unable to perceive the connexion between a cheap coat or a cheap tool, and the machines which make them plentiful, and therefore cheap. But even they, when the saving of labour by a machine is a saving which immediately affects them, are not slow to acknowledge the benefits they derive from that best of economy. The Scriptures allude to the painful condition of the " hewers of wood" and the " drawers of water ;" and certainly, in a state of society where there are no machines at all, or very rude machines, to cut down a tree and cleave it into logs, and to raise a bucket from a well, are very laborious occupations, the existence of which, to any extent, amongst a people, would mark them as remaining in a wretched condition. In our own country, at the present day, there are not many houses, in situations where water is at hand, that have not the windlass, or what is better, the pump, to raise this great necessary of life from the well. Some cottagers, however, have no such machines, and bitterly do they lament the want of them. We once met an old woman in a country district, tottering under the weight of a bucket, which she was labouring to carry up a hill. We asked her how she and her family were off in the world. She replied, that she could do pretty well with them, for they could all work, if it were not for one thing—it was one person's labour to fetch water from the spring; but, said she, if we had a pump handy, we should not have much to complain of. This old woman very wisely had no love of labour for its own sake; she saw no advantage in the labour of one of her family being given for the attainment of a good, which she knew might be attained by a very common invention. She wanted a machine to save that labour. Such a machine would have set at liberty a certain quantity of labour which was previously employed unprofitably; in other words, it would have left her or her children more time for more profitable work, and then the family earnings would have been increased.

But there is another point of view in which this machine would have benefited the good woman and her family. Water is not only necessary to drink and to prepare food with, but it is necessary for cleanliness, and cleanliness is necessary for health. If there is a scarcity of water, or if it requires a great deal of labour to obtain it, (which comes to the same thing as a scarcity,) the uses of water for cleanliness will be wholly or in part neglected. If the neglect becomes a habit, which it is sure to do, disease, and that of the worst sort, cannot be prevented.

When men gather together in large bodies, and inhabit towns or cities, a plentiful supply of water is the first thing to which they direct their attention. If towns are built in situations where pure water cannot be readily obtained, the inhabitants, and especially the poorer sort, suffer even more misery than results from the want of bread or clothes. In some cities of Spain, for instance, where the people understand very little about machinery, water, at particular periods of the year, is as dear as wine; and the labouring classes are consequently in a most miserable condition. In London, on the contrary, water is so plentiful, that between thirty and forty millions of gallons are daily supplied to the inhabitants; which quantity is distributed to about two hundred thousand houses and other buildings, the supply to each varying from one hundred to three hundred and fifty gallons daily. To many of the houses this water is, by the aid of machinery, not only delivered to the kitchens and wash-houses on the ground-floors, where it is most wanted, but is sent up to the very tops of the houses, to save even the comparatively little labour of fetching it from the bottom. All this is done at an average cost to each house of from less than a penny to about twopence a day; which is a less price than the labour of an able-bodied man would be worth to fetch a single bucket from a spring half a mile from his own dwelling.

And how did the inhabitants of London set about getting this great supply of water? How did they get a sufficient quantity, not only to use as much as they please for drinking, for cooking, and for washing, but obtained such an abundance, that the poorest man can afford to throw it away as if it cost nothing, into the channels which are also provided for carrying it off, and thus to free his own room or house from every impurity; and by so doing to render this vast place one of the most healthful cities in the world? They set about doing this great work by machinery; and they began to do it when the value of machinery in other things was not so well understood as it is now. As long ago as the year 1236, when a great want of water was felt in London, the little springs being blocked up and covered over by buildings, the ruling men of the city caused water to be brought from Tyburn, which was then a distant village, by means of pipes; and they laid a tax upon particular branches of trade to pay the expense of this great blessing to all. In succeeding times more pipes and conduits, that is, more machinery, was established for the same good purpose; and two centuries afterwards, King Henry the Sixth gave his aid to the same sort of works, in granting particular advantages in obtaining lead for making the pipes. The reason for this aid to such works was, as the royal decree set forth, that they were " for the common utility and decency of all the city, and *for the universal advantage*," and a very true reason this was.

As this great town more and more increased, more water-
works were found necessary ; till at last, in the reign of
James the First, which was nearly two hundred years
after that of Henry the Sixth, a most ingenious and en-
terprising man, and a great benefactor to his country,
Hugh Myddleton, undertook to bring a river of pure
water above thirty-eight miles out of its natural course,
for the supply of London. He persevered in this immense undertaking, in spite of every difficulty, till he at
last accomplished that great good which he had proposed,
of bringing wholesome water to every man's door. At
the present time, the New River, which was the work
of Hugh Myddleton, supplies seventeen millions of gallons of water every day ; and though the original pro-
jector was ruined by the undertaking, in consequence of
the difficulty which he had in procuring proper support,
such is now the general advantage of the benefit which
he procured for his fellow-citizens, and so desirous are
the people to possess that advantage, that a share in the
New River Company, which was at first sold at one
hundred pounds, is now worth fifteen thousand pounds.

Before the people of London had water brought to
their own doors, and even into their very houses, and
into every room of their houses where it is desirable to
bring it, they were obliged to send for this great article
of life—first, to the few springs which were found in the
city and its neighbourhood, and secondly, to the conduits
and fountains, which were imperfect mechanical contrivances for bringing it. The service-pipes to each
house are more perfect mechanical contrivances ; but they
could not have been rendered so perfect without engines,
which force the water above the level of the source from
which it is taken. When the inhabitants fetched their
water from the springs and conduits there was a great
deal of human labour employed ; and as in every large
community there are always people ready to perform
labour for money, many persons obtained a living by
carrying water. When the New River had been dug,
and the pipes had been laid down, and the engines had
been set up, it is perfectly clear that there would have

been no further need for these water-carriers. When the people of London could obtain two hundred gallons of water for twopence, they would not employ a man to fetch a single bucket from the river or the fountain at the same price. They would not, for the mere love of employing human labour directly, continue to buy an article very dear, which, by mechanical aid, they could buy very cheap. If they had resolved, from any mistaken notions about machinery, to continue to employ the water-carriers, they must have been contented with one gallon of water a day instead of two hundred gallons. Or if they had consumed a larger quantity, and continued to pay the price of bringing it to them by hand, they must have denied themselves other necessaries and comforts. They must have gone without a certain portion of food, or clothing, or fuel, which they are now enabled to obtain by the saving in the article of water. To have had for each house two hundred gallons of water, and in having this two hundred gallons of water, to have had the cleanliness and health which result from its use, would have been utterly impossible. At twopence a gallon, which would not have been a large price considering the distances to which it must have been carried, the same supply of water would have cost about nine millions of pounds sterling a year, and would have employed, at the wages of two shillings a day, more than one-half of all the present inhabitants of London, or eight hundred thousand people, that is, about four times the number of able-bodied men altogether contained in the metropolis. Such a supply therefore would have been utterly out of the question. To have supplied one gallon instead of two hundred gallons to each house at the same rate of wages, would have required the labour of twelve thousand men. It is evident that even this number could not have been employed in such an office; because had there been no means of supplying London with water but the means of human hands, London could not have increased to one-twentieth of its present size; there would not have been one-twentieth part of the population to have been supplied, and therefore six hundred water-carriers would have been an ample proportion to this population.

There is now, certainly, no labour to be performed by water-carriers. But suppose that five hundred years ago, when there were a small number of persons who gained their living by such drudgery, they had determined to prevent the bringing of water by pipes into London. Suppose also that they had succeeded; and that up to the present day we had no pipes or other mechanical aids for supplying the water. It is quite evident that if this misfortune had happened—if the welfare of the many had been retarded (for it never could have been finally stopped) by the ignorance of the few—London, as we have already shown, would not have had a twentieth part of its present population; and the population of every other town, depending as population does upon the increase of *profitable* labour, could never have gone forward. How then would the case have stood as to the amount of labour engaged in the supply of water? A few hundred, at the utmost a few thousand, carriers of water would have been employed throughout the kingdom; while the smelters and founders of iron of which water-pipes are made, the labourers who lay down these pipes, the founders of lead who make the service-pipes, and the plumbers who apply them; the carriers, whether by water or land, who are engaged in bringing them to the towns, the manufacturers of the engines which raise the water, the builders of the houses in which the engines stand,—these, and many other labourers and mechanics who directly and indirectly contribute to the same public advantage, could never have been called into employment. To have continued to use the power of the water-carriers, would have rendered the commodity two hundred times dearer than it is supplied by mechanical power. The present cheapness of production, by mechanical power, supplies employment to an infinitely greater number of persons than could have been required by a perseverance in the rude and wasteful system which belonged to former ages of ignorance and wretchedness.

CHAPTER XVIII.

THERE was a time when the people of England were very inferior to those of the Low Countries, of France, and of Germany, in various productions of manufacturing industry. We first gave an impulse to our woollen trade, which for several centuries was the great staple of the country, by procuring foreign workmen to teach our people their craft. Before that period the nations on the Continent had a proverb against us. They said, "the stranger buys of the Englishman the skin of the fox for a groat, and sells him the tail again for a shilling." The proverb meant, that we had not skill to convert the raw material into an article of use, and that we paid a large price for the labour and ingenuity which made our native material available to ourselves.

But still our intercourse, such as it was then, with "the stranger" was better than no intercourse. We gave the rough and stinking fox's skin for a groat, and we got the nicely dressed tippet for a shilling. The next best thing to dressing the skin ourselves, was to pay other people for dressing it. Without foreign communication we should not have got that article of clothing at all.

All nations that have made any considerable advance in civilization have been commercial nations. The arts of life are very imperfectly understood in countries which have little communication with the rest of the world, and consequently the inhabitants are poor and wretched ;— their condition is not bettered by the exchange with other countries, either of goods or of knowledge. They have the fox's skin, but they do not know how to convert it into value, by being furriers themselves, or by communication with "stranger" furriers.

The people of the East, amongst whom a certain degree of civilization has existed from high antiquity, were not only the growers of many productions which were unsuited to the climate and soil of Europe, but they were the manufacturers also. Cotton, for instance, was cultivated from time immemorial in Hindostan, in China, in Persia, and in Egypt. Cotton was a material easily grown and collected; and the patient industry of the people by whom it was cultivated, their simple habits, and their few wants, enabled them to send into Europe their manufactured stuffs of a fine and durable quality, under every disadvantage of land-carriage, even from the time of the ancient Greeks. Before the discovery, however, of the passage to India by the Cape of Good Hope, cotton goods in Europe were articles of great price and luxury. A French writer, M. Say, well observes, that although cotton stuffs were cheaper than silk (which was formerly sold for its weight in gold), they were still articles which could only be purchased by the most opulent; and that, if a Grecian lady could awake from her sleep of two thousand years, her astonishment would be unbounded, to see a simple country girl clothed with a gown of printed cotton, a muslin kerchief, and a coloured shawl.

When India was open to the ships of Europe, the Portuguese, the Dutch, and the English sold cotton goods in every market, in considerable quantities. These stuffs bore their Indian names of calicos and muslins; and whether bleached or dyed, were equally valued as amongst the most useful and ornamental articles of European dress.

In the seventeenth century France began to manufacture into stuffs the *raw* cotton imported from India, as Italy had done a century before. A cruel act of despotism drove the best French workmen, who were Protestants, into England, and we learned the manufacture. The same act of despotism, the revocation of the Edict of

Nantes, caused the settlement of silk-manufacturers in Spitalfields. We did not make any considerable progress in the art, nor did we use the material of cotton exclusively in making up the goods. The warp, or longitudinal threads of the cloth, were of flax, the weft only was of cotton ; for we could not twist it hard enough by hand to serve both purposes. This weft was spun entirely by hand with a distaff and spindle,—the same tedious process which prevails amongst the natives of India. Our manufacture, in spite of all these disadvantages, continued to increase ; so that about 1760, although there were fifty thousand spindles at work in Lancashire alone, the weaver found the greatest difficulty in procuring a sufficient supply of thread. Neither weaving nor spinning was then carried on in large factories. They were domestic occupations. The women of a family worked at the distaff or the hand-wheel, and there were two operations necessary in this department ; roving, or coarse spinning, reduced the carded cotton to the thickness of a quill, and the spinner afterwards drew out and twisted the roving into weft fine enough for the weaver. A writer on the cotton manufacture, Mr. Guest, states that very few weavers could procure weft enough to keep themselves constantly employed. " It was no uncommon thing," he says, " for a weaver to walk three or four miles in a morning, and call on five or six spinners, before he could collect weft to serve him for the remainder of the day ; and when he wished to weave a piece in a shorter time than usual, a new ribbon or gown was necessary to quicken the exertions of the spinner."

That the manufacture should have flourished in England at all under these difficulties is honourable to the industry of our country ; for the machinery used in weaving was also of the rudest sort, so that if the web was more than three feet wide, the labour of two men was necessary to throw the shuttle. English cotton goods, of course, were very dear, and there was little variety in them. The cloth made of flax and cotton was called fustian. We still received the calicos and printed cottons from India.

In a country like ours, where men have learned to think, and where ingenuity therefore is at work, a deficiency in material or in labour to meet the demand of a market is sure to call forth invention. It is nearly a century ago since it was perceived that spinning by machinery might give the supply which human labour was inadequate to produce, because, doubtless, the remuneration for that labour was very small. The work of the distaff, as it was carried on at that period, in districts partly agricultural and partly commercial, was, generally, an employment for the spare hours of the young women, and the easy industry of the old. It was a labour that was to assist in maintaining the family,—not a complete means for their maintenance. The supply of yarn was therefore insufficient, and ingenious men applied themselves to remedy that insufficiency. Spinning-mills were built at Northampton in 1733, in which, it is said, although we have no precise account of it, that an apparatus for spinning was erected. A Mr. Lawrence Earnshaw, of Mottram, in Cheshire, is recorded to have invented a machine, in 1753, to spin and reel cotton at one operation ; which he showed to his neighbours and then destroyed it, through the generous apprehension that he might deprive the poor of bread. We must admire the motive of this good man, although we are now enabled to show that his judgment was mistaken. Richard Arkwright, a barber of Preston, invented, in 1769, the principal part of the machinery for spinning cotton, and by so doing he gave bread to about two millions of people instead of fifty thousand ; and, assisted by subsequent inventions, raised the importation of cotton wool from less than two millions of pounds per annum, to above five hundred of millions ;—set in motion upwards of twelve millions of spindles, instead of fifty thousand ;—and increased the annual produce of the manufacture from two hundred thousand pounds sterling to thirty-four million pounds.

And how did he effect this great revolution ? He asked himself whether it was not possible, instead of a wheel which spins a single thread of cotton at a time, and by means of which the spinner could obtain in twenty-four hours about two ounces of thread,—whether it might not be possible to spin the same material upon a great number of wheels, from which many hundreds of threads might issue at the same moment. The difficulty was in giving to these numerous wheels, spinning so many threads, the peculiar action of two hands when they pinch, at a little distance from each other, a lock of cotton, rendering it finer as it is drawn out. It was necessary, also, at the same time to imitate the action of the spindle, which twisted together the filaments at the moment they had attained the necessary degree of fineness. It would be extremely difficult, if not impossible, to give an adequate idea, by words, of the complex machinery by which Arkwright accomplished his object, or of the subsequent improvements which have been made upon that machinery. But it may be desirable to describe that chief portion of his invention which enabled rollers to do the work of human fingers, with much greater precision, and incomparably cheaper.

We must suppose that by the previous operation of carding, the cotton wool has been so combed and prepared as to be formed into a long untwisted line of about the thickness of a man's finger. This line so formed (after it has been introduced into the machine we are about to describe) is called a *roving*, the old name in hand-spinning.

In order to convert this roving into a thread, it is necessary that the fibres, which are for the most part curled up, and which lie in all directions, should be stretched out and laid lengthwise, side by side; that they should be pressed together so as to give them a more compact form; and that they should be twisted, so as to unite them all firmly together. In the original method of spinning by the distaff, those operations were performed by the finger and thumb, and they were afterwards effected with greater rapidity, but less perfectly, by means of the long wheel and spindle. For the same purpose, Arkwright employed two pairs of small rollers, the one pair being placed at a little distance in front of the other. The lower roller in each pair is furrowed or fluted lengthwise, and the upper one is covered with leather; so that, as they revolve in contact with each other, they take fast hold of the cotton which passes between them. Both pairs of rollers are turned by machinery, which is so contrived that the second pair shall turn round with much more swiftness than the first. Now suppose that a roving is put between the first pair of rollers. The immediate effect is merely to press it together into a more compact form. But the roving has but just passed through the first pair of rollers, when it is received between the second pair; and as the rollers of the second pair revolve with greater velocity than those of the first, they draw the roving forwards with greater rapidity than it is given out by the first pair. Consequently, the roving will be lengthened in passing from one pair to the other; and the fibres of which it is composed will be drawn out and laid lengthwise side by side. The increase of length will be exactly in proportion to the increased velocity of the second pair of rollers.

Two or more rovings are generally united in this operation. Thus, suppose that two rovings are introduced together between the first pair of rollers, and that the second pair of rollers moves with twice the velocity of the first. The new roving thus formed by the union of the two, will then be of exactly twice the length of either of the original ones. It will therefore contain exactly the same quantity of cotton per yard. But its parts will be very differently arranged, and its fibres will be drawn out longitudinally, and will be thus much better fitted for forming a thread. This operation of doubling and drawing is repeated as often as is found necessary, and the requisite degree of twist is given by a machine similar to the spindle and fly of the common flax-wheel.

The fineness with which the cotton thread can be drawn out, by this machinery, may be gathered from the fact, that Mr. John Pollard, of Manchester, spun, in 1792, on the mule (the name of a particular description of the cotton-spinning machinery), no fewer than two hundred and seventy-eight hanks of yarn, forming a thread upwards of one hundred and thirty-two miles in length, from a single pound of raw cotton. Of the rapidity with which some portions of the machinery work, we may form an idea from the fact that the very finest thread which is used in making lace is passed through the strong flame of a lamp, which burns off the fibres, without burning the thread itself. The velocity with which the thread moves is so great, that we cannot perceive any motion at all. The line of thread, passing off a wheel through the flame, looks as if it were perfectly at rest; and it appears a miracle that it is not burnt.

The invention of Arkwright—the substitution of rollers for fingers—changed the commerce of the world. The machinery by which a man, or woman, or even a child could produce two hundred threads where one was produced before, caused a cheapness of production much greater than that of India, where human labour is scarcely worth anything. But the fabric of cotton was also infinitely improved by the machinery. The hand of the spinner was unequal in its operations. It sometimes produced a fine thread, and sometimes a coarse one; and therefore the quality of the cloth could not be relied upon. The yarn which is spun by machinery is sorted with the greatest exactness, and numbered according to its quality. This circumstance alone, which could only result from machinery, has a direct tendency to diminish the cost of production. Machinery not only adds to human power, and economizes human time, but it works up the most common materials into articles of value, and equalizes the use of valuable materials. Thus, in linen of which the thread is spun by the hand, a thick thread and a thin thread will be found side by side; and, therefore, not only is material wasted, but the fabric is less durable, because it wears unequally.

These circumstances—the diminished cost of cotton goods, and the added value to the quality—have rendered it impossible for the cheap labour of India to come into the market against the machinery of Europe. The trade in Indian cotton goods is gone for ever. Not even the caprices of fashion can have an excuse for purchasing the dearer commodity. We make it cheaper, and we make it better. The trade in cotton, as it exists in the present day, is the great triumph of human ingenuity. We bring the raw material from the country of the people who grow it, on the other side of our globe; we manufacture it by our machines into articles which we used to buy from them ready-made; and taking back those articles to their own markets, encumbered with the cost of transport for fourteen thousand miles, and encumbered also with the taxes which the State has laid upon it in many various ways, we sell it to these very people cheaper than they can produce it themselves, and they buy it therefore with eagerness.

CHAPTER XIX.

NEARLY twenty years after Arkwright had begun to spin by machinery, the price of a particular sort of cotton yarn much used in the manufacture of calico was thirty-eight shillings a pound. That same yarn is now sold for between three and four shillings, or one-twelfth of its price fifty years ago. If cotton goods were worn only by the few rich, as they were worn in ancient times, and even in the latter half of the last century, that difference of price would not be a great object; but the price is a very important object when every man, woman, and child in the United Kingdom has to pay it. The seven hundred million yards of cloth which are annually retained for home consumption, distributed amongst twenty-seven millions of population, allows twenty-six yards every year for each individual. We will suppose that no individual would buy these twenty-six yards of cloth unless he or she wanted them; that this plenty of cloth is a desirable thing; that it is conducive to warmth and cleanliness, and therefore to health; that it would be a great privation to go without the cloth. At five-pence a yard, the seven hundred million yards of cloth amount to above fourteen million pounds sterling. At half-a-crown a yard, which we will take as the average price about thirty or forty years ago, they would amount to eighty-seven millions of pounds sterling—an amount equal to all the taxes annually paid in Great Britain and Ireland. At twelve or fourteen times the present price, or six shillings a yard, which proportion we get by knowing the price of yarn forty years ago and at the present day, the cost of seven hundred million yards of cotton cloth would be one hundred and seventy-five millions of pounds sterling. It is perfectly clear that no such sum of money could be paid for cotton goods, and that in fact instead of between fourteen and fifteen millions being spent in this article of clothing by persons of all classes, in consequence of the cheapness of the commodity, we should go back to very nearly the same consumption that existed before Arkwright's invention, that is, to the consumption of the year 1750, when the whole amount of the cotton manufacture of the kingdom did not exceed the annual value of two hundred thousand pounds. At that rate of value, the quantity of cloth manufactured could not have been equal to one five-hundredth part of that which is now manufactured for home consumption. Where one person eighty years ago consumed one yard, the consumption per head has risen to about twenty-six yards. We ask, therefore, if this vast difference in the comforts of every family, by the ability which they now possess of easily acquiring warm and healthful clothing, is not a clear gain to all society, and to every one as a portion of society ? It is more especially a gain to the females and the children of your families, whose condition is always degraded when clothing is scanty. The power of procuring cheap clothing for themselves, and for their children, has a tendency to raise the condition of females more than any other addition to their stock of comfort. It cultivates habits of cleanliness and decency ; and those are little acquainted with the human character who can doubt whether cleanliness and decency are not only great aids to virtue, but virtues themselves. John Wesley said that cleanliness was next to godliness. There is little self-respect amidst dirt and rags, and without self-respect there can be no foundation for those qualities which most contribute to the good of society. The power of procuring useful clothing at a cheap price has raised the condition of women amongst us, and the influence of the condition of women upon the welfare of a community can never be too highly estimated.

That the manufacture of cotton by machinery has produced one of the great results for which machinery is to be desired, namely, cheapness of production, cannot, we think, be doubted. If increased employment of human labour has gone along with that cheapness of production, even the most prejudiced can have no doubt of the advantages of this machinery to all classes of the community.

At the time that Arkwright commenced his machinery, a man named Hargreave, who had set up a less perfect invention, was driven out of Lancashire, at the peril of his life, by a combination of the old spinners by the wheel. In 1789, when the spinning machinery was introduced into Normandy, the hand-spinners there also destroyed the mills, and put down the manufacture for a time. Lancashire and Normandy are now, in England and France, the great seats of the cotton manufacture. The people of Lancashire and Normandy had not formerly the means, as we have now, of knowing that cheap production produces increased employment. There were many examples of this principle formerly to be found in arts and manufactures; but the people were badly educated upon such subjects, principally because studious and inquiring men had thought such matters beneath their attention. We live in times more favourable for these researches. The people of Lancashire and Normandy, at the period we mention, being ignorant of what would conduce to their real welfare, put down the machines. In both countries they were a very small portion of the community that attempted such an illegal act. The weavers were interested in getting cotton yarn cheap, so the combination was opposed to their interests; and the spinners were chiefly old women and girls, very few in number, and of little influence. Yet they and their friends, both in England and France, made a violent clamour; and but for the protection of the laws, the manufactories in each country would never have been set up. What was the effect upon the condition of this very population? M. Say, in his ' Complete Course of Political Economy,' states, upon the authority of an English manufacturer of fifty years' experience, that, in

H

ten years after the introduction of the machines, the
people employed in the trade, spinners and weavers,
were more than forty times as many as when the spinning was done by hand. It was calculated, in 1825, and
since that year, that the power of twenty thousand horses
was employed in the spinning of cotton ; and that the
power of each horse yielded, with the aid of machinery,
as much yarn as one thousand and sixty-six persons
could produce by hand. If this calculation be correct,
and there is no reason to doubt it, the spinning machinery of Lancashire alone produced, in 1825, as much
yarn as would have required twenty-one million three
hundred and twenty thousand persons to produce with
the distaff and spindle. This immense power, which is
nearly equal to the population of the United Kingdom,
might be supposed to have superseded human labour
altogether in the production of cotton yarn. It did no
such thing. It gave a new direction to the labour that
was formerly employed at the distaff and spindle ; but it
increased the quantity of labour altogether employed in
the manufacture of cotton, at least a hundred fold. It
increased it too where an increase of labour was most desirable. It gave constant, easy, and not unpleasant
occupation to women and children. In all the depart-
ments of cotton spinning, and in many of those of weaving by the power-loom, women and children are employed. There are degrees, of course, in the agreeable
nature of the employment, particularly as to its being
more or less cleanly. But there are extensive apart-
ments in large cotton-factories, where great numbers of
females are daily engaged in processes which would not
soil the nicest fingers, dressed with the greatest neatness,
and clothed in materials (as all women are now clothed)
that were set apart for the highest in the land a century
ago. And yet there are some who regret that the aged
crones no longer sit in the cottage chimney, earning a
few pence daily by their rude industry at the wheel!

The creation of employment amongst ourselves by the
cheapness of cotton goods produced by machinery, is not
to be considered as a mere change from the labour of

India to the labour of England. It is a creation of employment, operating just in the same manner as the machinery did for printing books. The Indian, it is true, no longer sends us his calicos and his coloured stuffs; we make them ourselves. But he sends us forty times the amount of raw cotton that he sent when the machinery was first set up. In 1781 we imported five million pounds of cotton wool. In 1843 we imported five hundred and thirty-one million pounds—enough to make twelve hundred and sixty million yards of cloth— which is about two yards a piece for every human being in the world. The workman on the banks of the Ganges (the great river of India) is no longer weaving calicos for us, in his loom of reeds under the shade of a mango tree; but he is gathering for us forty times as much cotton as he gathered before, and making forty times as much indigo for us to colour it with. The export of cotton has made such a demand upon the Indian power of labour, that even the people of Hindostan, adopting European contrivances, have introduced machinery to pack the cotton. Bishop Heber says, that he was frequently interested by seeing, at Bombay, immense bales of cotton lying on the piers, and the ingenious screw, by which an astonishing quantity is pressed into the canvas bags. The Chinese, on the contrary, from the want of these contrivances to press the cotton so close in bags, sell their cotton to us at much less profit; for they pack it so loosely, that it occupies three times the bulk of the Indian cotton, and the freight costs twelve times the price on this account. When the Chinese acquire the knowledge from other nations, which recent events render probable, they will know the value of mechanical skill, in preference to unassisted manual labour.

The arguments for the use of machinery, that may be derived from the manufacture of SILK, are precisely the same as those we have exhibited in the manufacture of cotton. The cost of production has been lessened—the employment of the producers has been increased. When the frame-work knitters of silk stockings petitioned

H 2

Oliver Cromwell for a charter, they said, " the English-man buys silk of the stranger for [twenty marks, and sells him the same again for one hundred pounds." The higher pride of the present day is that we buy four million pounds of raw silk from the stranger, employ a large number of our own people in the manufacture of it by the aid of machinery, and sell it to the stranger, and our own people, at a price as low as that of the calico of half a century ago.

The manufactures of WOOLLEN CLOTH, and of LINEN CLOTH, partly carried on with materials produced by ourselves, and partly with wool and flax bought from other nations, have increased, with the use of machinery, in the same way as the cotton manufacture. In both cases, the article produced is diminished in price.

CHAPTER XX.

THE beaver builds his huts with the tools which nature has given him. He gnaws pieces of wood in two with his sharp teeth, so sharp, that the teeth of a similar animal, the agouti, form the only cutting-tool which some rude nations possess. When the beavers desire to move a large piece of wood, they join in a body to drag it along.

Man has not teeth that will cut wood. But he has reason, which directs him to the choice of much more perfect tools.

Some of the great monuments of antiquity, such as the pyramids of Egypt, are constructed of enormous blocks of stone brought from distant quarries. We have no means of estimating, with any accuracy, the mechanical knowledge possessed by the people engaged in these works. It was, probably, very small, and consequently, the human labour employed in such edifices was not only enormous in quantity, but exceedingly painful to the workmen. The Egyptians, according to Herodotus, a Greek writer who lived two thousand five hundred years ago, hated the memory of the kings who built the pyramids. He tells us that the great pyramid occupied a hundred thousand men for twenty years in its erection, without counting the workmen who were employed in hewing the stones, and in conveying them to the spot where the pyramid was built. Herodotus speaks of this work as a torment to the people; and doubtless, the labour engaged in raising huge masses of stone, that was extensive enough to employ a hundred thousand men

for twenty years, which is equal to two million of men
for one year, must have been fearfully tormenting without machinery, or with very imperfect machinery.   It
has been calculated that the steam-engines of England,
worked by thirty-six thousand men, would raise the same
quantity of stones from the quarry, and elevate them to
the same height as the great pyramid, in the short time
of eighteen hours.   The people of Egypt groaned for
twenty years under this enormous work.   The labourers
groaned because they were sorely tasked; and the rest
of the people groaned because they had to pay the
labourers.   The labourers lived, it is true, upon the
wages of their labour, that is, they were paid in food—
kept like horses—as the reward of their work.   Herodotus says, that it was recorded on the pyramid, that the
onions, radishes, and garlic which the labourers consumed, cost sixteen hundred talents of silver: an immense sum, equivalent to several million pounds.   But
the onions, radishes, and garlic, the bread, and clothes
of the labourer, were wrung out of the profitable labour
of the rest of the people.   The building of the pyramid
was an unprofitable labour.   There was no immediate or
future source of produce in the pyramid; it produced
neither food, nor fuel, nor clothes, nor any other necessary.   The labour of a hundred thousand men for twenty
years, stupidly employed upon this monument, without
an object beyond that of gratifying the pride of the
tyrant who raised it, was a direct tax upon the profitable
labour of the rest of the people.

> " Instead of useful works, like nature great,
> Enormous cruel wonders crush'd the land."

But admitting that it is sometimes desirable for nations
and governments to erect monuments which are not of
direct utility,—which may have an indirect utility in re-
cording the memory of great exploits, or in producing
feelings of reverence or devotion,—it is clearly an advantage that these works, as well as all other works,
should be performed in the cheapest manner; that is,
that human labour should derive every possible assistance

from mechanical aid. We will give an illustration of the differences of the application of mechanical aid in one of the first operations of building, the moving a block of stone. The following statements are the result of actual experiment upon a stone weighing ten hundred and eighty pounds.

To drag this stone along the smoothed floor of the quarry required a force equal to seven hundred and fifty-eight pounds. The same stone dragged over a floor of planks required six hundred and fifty-two pounds. The same stone placed on a platform of wood, and dragged over the same floor of planks, required six hundred and six pounds. When the two surfaces of wood were soaped as they slid over each other, the force required to drag the stone was reduced to one hundred and eighty-two pounds. When the same stone was placed upon rollers three inches in diameter, it required, to put it in motion along the floor of the quarry, a force only of thirty-four pounds; and by the same rollers upon a wooden floor, a force only of twenty-eight pounds. You will see, therefore, that without any mechanical aid, it would require the force of four or five men to set that stone in motion. With the mechanical aid of two surfaces of wood soaped, the same weight might be moved by one man. With the more perfect mechanical aid of rollers, the same weight might be moved by a very little child.

From these statements it must be evident that the cost of a block of stone very much depends upon the quantity of labour necessary to move it from the quarry to the place where it is wanted to be used. We have seen that with the simplest mechanical aid labour may be reduced sixty-fold. With more perfect mechanical aid, such as that of water-carriage, the labour may be reduced infinitely lower. Thus, the streets of London are paved with granite from Scotland at a moderate expense.

The cost of timber, which enters so largely into the cost of a house, is, in a great degree, the cost of transport. We load two thousand ships yearly with the timber which we import from the Baltic Sea, and from North America.

In countries where there are great forests, timber-trees are worth nothing where they grow, except there are ready means of transport. In many parts of North America, the great difficulty which the people find is in clearing the land of the timber. The finest trees are not only worthless, but are a positive encumbrance, except when they are growing upon the banks of a great river; in which case the logs are thrown into the water, or formed into rafts, being floated several hundred miles at scarcely any expense. The same stream which carries them to a seaport turns a mill to saw the logs into planks; and when sawn into planks the timber is put on shipboard, and carried to distant countries where timber is wanted. Thus mechanical aid alone gives a value to the timber, and by so doing employs human labour. The stream that floats the tree, the sawing-mill that cuts it, the ship that carries it across the sea, enable men profitably to employ themselves in working it. Without the stream, the mill, and the ship, those men would have no labour, because none could afford to bring the timber to their own doors.

We build in this country more of brick than of stone, because brick earth is found almost everywhere, and stone fit for building is found only in particular districts. Bricks pay to the State a duty of five shillings and tenpence a thousand; and yet at the kilns they may be bought under forty shillings a thousand, which is less than a halfpenny a piece. How is it that bricks are so cheap? Because they are made by machinery. The clay is ground in a horse-mill; the wooden mould in which every brick is made singly, is a copying machine. One brick is exactly like another brick. Every brick is of the form of the mould in which it is made. Without the mould the workman could not make the brick of uniform dimensions; and without this uniformity the after labour of putting the bricks together would be greatly increased. Without the mould the workman could not form the bricks quickly; his own labour would be increased tenfold. Because bricks are cheap, one thousand five hundred million bricks, as the Excise returns show, are made in England in a year;

and thus the simple machine of the mould not only gives employment to a great many brick-makers who would not be employed at all, but also to a great many bricklayers who would also want employment, if the original cost of production were so enormously increased.

What an infinite variety of machines, in combination with the human hand, is found in a carpenter's chest of tools! The skilful hand of the workman is the *power* which sets these machines in motion ; just as the wind or the water is the power of a mill, or the clastic force of vapour the power of a steam-engine. When Mr. Boulton, the partner of the celebrated James Watt, waited upon George III. to explain one of the great improvements of the steam-engine which they had effected, the king said to him, " What do you sell, Mr. Boulton ?" and the honest engineer answered, " What kings, sire, are all fond of —*power*." There are people at Birmingham who let out *power*, that is, there are people who have steam-engines who will lend the use of them, by the day or the hour, to persons who require that saving of labour in their various trades ; so that a person who wants the strength of a horse, or half a horse, to turn a wheel for grinding, or for setting a lathe in motion, hires a room, or part of a room, in a mill, and has just as much power as he requires. The *power* of a carpenter is in his hand, and the machines moved by that power are in his chest of tools. Every tool which he possesses has for its object to reduce labour, to save material, and to ensure accuracy —the objects of all machines. What a quantity of waste both of time and stuff is saved by his foot-rule ! and when he chalks a bit of string and stretches it from one end of a plank to the other, to jerk off the chalk from the string, and thus produce an unerring line upon the face of the plank, he makes a little machine, which saves him great labour. Every one of his hundreds of tools, capable of application to a vast variety of purposes, is an invention to save labour. Without some tool the carpenter's work could not be done at all by the human hand. A knife would do very laboriously what is done very quickly by a hatchet. The labour of using a hatchet, and the material which it wastes, are saved twenty times over by the saw. But when the more delicate operations of carpentry are required—when the workman uses his planes, his rabbet-planes, his fillisters, his bevels, and his centre-bits—what an infinitely greater quantity of labour is economized, and how beautifully that work is performed, which, without them, would be rough and imperfect! Every boy of mechanical ingenuity has tried with his knife to make a boat; and with a knife only it is the work of weeks. Give him a chisel, and a gouge, and a vice to hold his wood, and the little boat is the work of a day. Let a boy try to make a round wooden box, with a lid, having only his knife, and he must be expert indeed to produce any thing that will be neat and serviceable. Give him a lathe and chisels, and he will learn to make a tidy box in half an hour.

If carpenters had not tools to make houses, there would be few houses made; and those that were made would be as rough as the hut of the savage who has no tools. The people would go without houses, and the carpenter would go without work,—to say nothing of the people, who would also go without work, that now make tools for the carpenter.

How great a variety of things are contained in an ironmonger's shop! Half his store consists of tools of one sort or another to save labour; and the other half consists of articles of convenience or elegance most perfectly adapted to every possible want of the builder or the maker of furniture. The uncivilized man is delighted when he obtains a nail,—any nail. A carpenter and joiner, who supply the wants of a highly civilized community, are not satisfied unless they have a choice of nails, from the finest brad to the largest clasp nail. A savage thinks a nail will hold two pieces of wood together more completely than anything else in the world. It is seldom, however, that he can afford to put it to such a use. If it is large enough, he makes it into a chisel. An English joiner knows that screws will do the work more perfectly in some cases than any nail; and therefore we have as great a variety of screws as of nails.

The commonest house built in England has hinges, and locks, and bolts. A great number are finished with ornamented knobs to door-handles, with bells and bell-pulls, and a thousand other things that have grown up into necessities, because they save domestic labour, and add to domestic comfort. And many of these things really are necessities. M. Say, a French writer, gives us an example of this; and as his story is an amusing one, besides having a moral, we may as well copy it:—

" Being in the country," says he, " I had an example of one of those small losses which a family is exposed to through negligence. For the want of a latchet of small value, the wicket of a barn-yard leading to the fields was often left open. Every one who went through drew the door to ; but as there was nothing to fasten the door with, it was always left flapping ; sometimes open, and sometimes shut. So the cocks and hens, and the chickens, got out, and were lost. One day a fine pig got out, and ran off into the woods ; and after the pig ran all the people about the place,—the gardener, and the cook, and the dairy-maid. The gardener first caught sight of the runaway, and, hastening after it, sprained his ankle ; in consequence of which the poor man was not able to get out of the house again for a fortnight. The cook found, when she came back from pursuing the pig, that the linen she had left by the fire had fallen down, and was burning ; and the dairy-maid having, in her haste, neglected to tie up the legs of one of her cows, the cow had kicked a colt, which was in the same stable, and broken its leg. The gardener's lost time was worth twenty crowns, to say nothing of the pain he suffered. The linen which was burned, and the colt which was spoiled, were worth as much more. Here, then, was caused a loss of forty crowns, as well as much trouble, plague, and vexation, for the want of a latch which would not have cost threepence." M. Say's story is one of the many examples of the truth of the old proverb— " for want of a nail the shoe was lost, for want of a shoe the horse was lost, for want of a horse the man was lost."

Nearly all the great variety of articles in an iron-monger's shop are made by machinery. Without machinery they could not be made at all, or they would be sold at a price which would prevent them being commonly used. With machinery, their manufacture employs large numbers of artisans, who would be otherwise unemployed. There are hundreds of ingenious men at Birmingham who go into business with a capital acquired by their savings as workmen, for the purpose of manufacturing some one single article used in finishing a house, such as the knob of a lock. All the heavy work of their trade is done by machinery. The cheapness of the article creates workmen; and the savings of the workmen accumulate capital to be expended in larger works, and to employ more workmen.

The furniture of a house, some may say—the chairs, and tables, and bedsteads—is made nearly altogether by hand. True. But tools are machines; and further, we owe it to what men generally call machinery, that such furniture, even in the house of a very poor man, is more tasteful in its construction, and of finer material, than that possessed by a nobleman a hundred years ago. How is this? Machinery (that is ships) has brought us much finer woods than we grow ourselves; and other machinery (the sawing-mill) has taught us how to render that fine wood very cheap, by economising the use of it. At a veneering-mill, that is, a mill which cuts a mahogany log into thin plates, much more delicately and truly, and in infinitely less time, than they could be cut by the hand, two hundred and forty square feet of mahogany are cut by one circular saw in an hour. A veneer, or thin plate, is cut off a piece of mahogany, six feet six inches long, by twelve inches wide, in twenty-five seconds. What is the consequence of this? A mahogany table is made almost as cheap as a deal one; and thus the humblest family in England may have some article of mahogany, if it be only a tea-caddy. And let it not be said that deal furniture would afford as much happiness: for a desire for comfort, and even for some degree of elegance, gives a refinement to the character, and, in a certain degree, raises our self-respect. Diogenes, who is said to have lived in a tub, was a great philosopher; but it is not necessary to live in a tub to be wise and virtuous. Nor is that the likeliest plan for becoming so. The probability is, that a man will be more wise and virtuous, in proportion as he strives to surround himself with the comforts and decent ornaments of his station.

We think that, with regard to buildings and the furniture of buildings, it will be admitted that machinery, in the largest sense of the word, has increased the means of every man to procure a shelter from the elements, and to give him a multitude of conveniences within that shelter. Most will agree, we think, that a greater number of persons are profitably employed in affording this shelter and these conveniences, with tools and machines, than if they possessed no such mechanical aids to their industry. When the account of the population of Great Britain was last taken, in 1841, there were three million four hundred and sixty-five thousand houses inhabited, and thirty thousand houses being built. In New Zealand, which is as large as Great Britain, there are not ten thousand native habitations; and their huts are made of the roughest materials, and in the most comfortless manner. The nation which has mechanical knowledge has two hundred and fifty times as many houses as the nation without these aids; and the poorest house of the civilized people is fifty times as commodious as the finest house of the uncivilized people. We cannot doubt which nation has the most employment for builders.

CHAPTER XXI.

WE have seen what machinery will do, in working up
valuable materials, such as cotton, brought at a large cost
from foreign countries, so as to sell the manufactured
goods at a rate which does not exclude even the poorest
from their purchase. Let us see what the same sort of
mechanical ingenuity can effect, in producing the most
useful and ornamental articles of domestic life, from the
common earth which may be had for digging. Without
chemical and mechanical skill we should neither have
glass nor pottery ; and without these articles, how much
lowered beneath his present station, in point of comfort
and convenience, would be the humblest peasant in the
land !

The cost of GLASS is almost wholly the wages of labour, as the materials are very abundant, and may be
said to cost almost nothing ; and glass is much more
easily worked than any other substance.

Hard and brittle as it is, it has only to be heated, and
any form that the workman pleases may be given to it.
It melts ; but when so hot as to be more susceptible of
form than wax or clay, or any thing else that we are
acquainted with, it still retains a degree of toughness and
capability of extension superior to that of many solids,
and of every liquid ; when it has become red hot all its
brittleness is gone, and a man may do with it just as he
pleases. He may press it into a mould ; he may take a
lump of it upon the end of an iron tube, and, by blowing
into the tube with his mouth, (keeping the glass hot all
the time,) he may swell it out into a hollow ball. He

may mould that ball into a bottle; he may draw it out lengthways into a pipe; he may cut it open into a cup; he may open it with shears, whirl it round with the edge in the fire, and thus make it into a circular plate. He may also roll it out into sheets, and spin it into threads as fine as a cobweb. In short, so that he keeps it hot, and away from substances by which it may be destroyed, he can do with it just as he pleases. All this, too, may be done, and is done with large quantities every day, in less time than any one would take to give an account of it. In the time that the readiest speaker and clearest describer were telling how one quart bottle is made, an ordinary set of workmen would make some dozens of bottles.

But though the materials of glass are among the cheapest of all materials, and the substance the most obedient to the hand of the workman, there is a great deal of knowledge necessary before glass can be made. It can be made profitably only at large manufactories, and those manufactories must be kept constantly at work night and day.

Glass does not exist in a natural form in many places. The sight of native crystal, probably, led men to think originally of producing a similar substance by art. The fabrication of glass is of high antiquity. The historians of China, Japan, and Tartary speak of glass manufactories existing there more than two thousand years ago. An Egyptian mummy two or three thousand years old, which was exhibited in London, was ornamented with little fragments of coloured glass. The writings of Seneca, a Roman author who lived about the time of our Saviour, and of St. Jerome, who lived five hundred years afterwards, speak of glass being used in windows. It is recorded that the Prior of the convent of Weymouth, in Dorsetshire, in the year 674, sent for French workmen to glaze the windows of his chapel. In the twelfth century the art of making glass was known in this country. Yet it is very doubtful whether glass was employed in windows, excepting those of churches and the houses of the very rich, for several centuries afterwards; and it is quite certain that the period is comparatively recent, as we have shown in Chap. VIII., when glass windows were used for excluding cold and admitting light in the houses of the great body of the people, or that glass vessels were to be found amongst their ordinary conveniences. The manufacture of glass in England now employs many thousand people, because the article, being cheap, is of universal use. The government has wisely taken off the duty on glass; and as the article becomes cheaper, so will the people employed in its manufacture become more numerous.

Machinery, as we commonly understand the term, is not much employed in the manufacture of glass; but chemistry, which saves as much labour as machinery, and performs work which no machinery could accomplish, is very largely employed. The materials of which glass is made are sand or earth, and vegetable matter, such as kelp or burnt sea-weed. These materials are put in a state of fusion by the heat of an immense furnace. It requires a red heat of sixty hours to prepare the material of a common bottle. Nearly all glass, except glass for mirrors, is what is called blown. The machinery is very simple, consisting only of an iron pipe, and the lungs of the workman; and the process is perfected in all its stages by great subdivision of labour, producing extreme neatness and quickness in all persons employed in it. For instance, a wine-glass is made thus:—One man (the blower) takes up the proper quantity of glass on his pipe, and blows it to the size wanted for the bowl; then he whirls it round on a reel, and draws out the stalk. Another man (the footer) blows a smaller and thicker ball, sticks it to the end of the stalk of the blower's glass, and breaks his pipe from it. The blower opens that ball, and whirls the whole round till the foot is formed. Then a boy dips a small rod in the glass-pot, and sticks it to the very centre of the foot. The blower, still turning the glass round, takes a bit of iron, wets it in his mouth, and touches the ball at the place where he wishes the mouth of the glass to be. The glass separates, and the boy takes it to the finisher, who turns the mouth of it; and by a peculiar

swing that he gives it round his head, makes it perfectly
circular, at the same time that it is so hardened as to be
easily snapped from the rod.   Lastly the boy takes it on
a forked iron to the annealing furnace, where it is cooled
gradually.

All these operations require the greatest nicety in the
workmen; and would take a long time in the performance,
and not be very neatly done after all, if they were all
done by one man.   But the quickness with which they
are done by the division of labour is perfectly wonderful.

The cheapness of glass for common use, which cheapness is produced by chemical knowledge and the division
of labour, has set the ingenuity of man to work to give
greater beauty to glass as an article of luxury.   The employment of sharp-grinding wheels put in motion by a
treadle, and used in conjunction with a very nice hand,
produces *cut* glass.   Cut glass is now comparatively so
cheap, that scarcely a family of the middle ranks is without some beautiful article of this manufacture.   And yet,
cheap as glass is, a great deal of even its present price is
tax.

If, to increase the mere labour of the glass-blowers, it
was resolved that furnaces of less power should be employed, or if, for the same purpose, the subdivision of
labour were abolished, and one man were to make a wine-
glass in all its stages, the working men of England would
have no glass in their windows.   If the glass-cutters were
to lay down their wheels, and take to files, the tradesman
would have no cut-glass decanter on his table.   The rich
only would possess cut-glass vessels of any beauty of construction ; and, consequently, the glass-cutters would
dwindle down from thousands to hundreds, and even to
tens.

There are two kinds of POTTERY—common potters'
ware, and porcelain of China.   The first is a pure kind
of brick ; and the second a mixture of very fine brick and
glass.   Almost all nations have some knowledge of pottery ; and those of the very hot countries are sometimes
satisfied with dishes formed by their fingers without any

tool, and dried by the heat of the sun. In England pottery of every sort, and in all countries good pottery, must be baked or burnt in a kiln of some kind or other.

Vessels for holding meat and drink are almost as indispensable as the meat and drink themselves; and the two qualities in them that are most valuable, are, that they shall be cheap, and easily cleaned. Pottery, as it is now produced in England, possesses both of these qualities in the very highest degree. A white basin, having all the useful properties of the most costly vessels, may be purchased for twopence at the door of any cottage in England. There are very few substances used in human food that have any effect upon these vessels; and it is only rinsing them in hot water, and wiping them with a cloth, and they are clean.

The making of an earthen bowl would be to a man who made a first attempt no easy matter. Let us see how it is done so that it can be carried two or three hundred miles and sold for twopence, leaving a profit to the maker, and the wholesale and retail dealer.

The common pottery is made of pure clay and pure flint. The flint is found only in the chalk counties, and the fine clays in Devonshire and Dorsetshire; so that the materials out of which the pottery is made have to be carried from the South of England to Staffordshire, where the potteries are situated.

The great advantage that Staffordshire possesses is abundance of coal to burn the ware and supply the engines that grind the materials.

The clay is worked in water by various machinery till it contains no single piece large enough to be visible to the eye. It is like cream in consistence. The flints are burned. They are first ground in a mill and then worked in water in the same manner as the clay, the large pieces being returned a second time to the mill.

When both are fine enough, one part of flint is mixed with five or six of clay; the whole is worked to a paste, after which it is kneaded either by the hands or a machine; and when the kneading is completed, it is ready for the potter.

He has a little wheel which lies horizontally. He lays a portion of clay on the centre of the wheel, puts one hand, or finger if the vessel is to be a small one, in the middle, and his other hand on the outside, and, as the wheel turns rapidly round, draws up a hollow vessel in an instant. With his hands, or with very simple tools, he brings it to the shape he wishes, cuts it from the wheel with a wire, and a boy carries it off. The potter makes vessel after vessel, as fast as they can be carried away.

They are partially dried; after which they are turned on a lathe and smoothed with a wet sponge when necessary.

Only round vessels can be made on the wheel; those of other shapes are made in moulds of plaster.

Handles and other solid parts are pressed in moulds, and stuck on while they and the vessels are still wet.

The vessels thus formed are first dried in a stove, and, when dry, burnt in a kiln. They are in this state called biscuit. If they are finished white, they are glazed by another process. If they are figured, the patterns are engraved on copper, and printed on coarse paper rubbed with soft soap. The ink is made of some colour that will stand the fire, ground with earthy matter. These patterns are moistened and applied to the porous biscuit, which absorbs the colour, and the paper is washed off, leaving the pattern on the biscuit.

The employment of machinery to do all the heavy part of the work, the division of labour, by which each workman acquires wonderful dexterity in his department, and the conducting of the whole upon a large scale, give bread to a vast number of people, make the pottery cheap, and enable it to be sold at a profit in almost every market in the world. It is not seventy years since the first pottery of a good quality was extensively made in England; and before that time what was used was imported, the common ware from Delf, in Holland, (from which it acquired its name,) and the porcelain from China. We now annually export fifty-three million pieces of earthenware to all parts of the world.

CHAPTER XXII.

If the facts which we have stated in the preceding chapters have been duly considered, it appears that we cannot much doubt that in articles of the most absolute necessity, machinery has at the same time diminished the cost of production, and added to the numbers of the workmen. Without machinery, as we have shown, it would be impossible to raise food, to manufacture implements, to supply fuel and water, to carry on communication, to produce clothes, to build houses and furnish them, and to distribute knowledge, either *at all*, or at least at a price which should allow all men, more or less, to partake these great blessings of civilization. In the present chapter we propose to show some very curious effects of machinery in the production of articles of inferior value, certainly, to those chief necessaries of life which we have mentioned, but which are in such general use amongst all of us, however trifling they may appear in themselves, that the want of them would be felt as a severe privation. Without machines they could not be made at all; or they would be made very coarsely, as mere curiosities. With machines they are made in such numbers that they constitute very large branches of trade, and give employment to hundreds of thousands of people, in assisting the machines, or in perfecting what they produce.

There is an article employed in dress, which is at once so necessary and so beautiful, that the highest lady in the land uses it, and yet so cheap, that the poorest peasant's wife is enabled to procure it. The quality of the article is as perfect as art can make it; and yet, from the enormous quantities consumed by the great mass of the people, it is made so cheap that the poor can purchase the best kind, as well as the rich. It is an article of universal use. United with machinery, many hundreds, and even thousands, are employed in making it. But if the machinery were to stop, and the article were made by human hands alone, it would become so dear that the richest only could afford to use it; and it would become, at the same time, so rough in its appearance, that those very rich would be ashamed of using it. The article we mean is a pin.

Machinery of all kinds is difficult to be described by words. It is not necessary for us to describe the machinery used in pin-making, to make the reader comprehend its effects. A pin is made of brass. We have seen how metal is obtained from ore by machinery, and therefore we will not go over that ground. But suppose the most skilful workman has a lump of brass ready by his side, to make it into pins with common tools,—with a hammer and with a file. He beats it upon an anvil, till it becomes nearly thin enough for his purpose. A very fine hammer, and a very fine touch, must he have to produce a pin of any sort,—even a large corking pin! But the pin made by machinery is a perfect cylinder. To make a metal, or even a wooden cylinder, of a considerable size, with files and polishing, is an operation so difficult, that it is never attempted; but with a lathe and a sliding rest it is done every hour by a great many workmen. How much more difficult would it be to make a perfect cylinder the size of a pin? A pin hammered out by hand would present a number of rough edges that would tear the clothes, as well as hold them together. It would not be much more useful or ornamental than the skewer of bone, with which the woman of the Sandwich Islands fastens her mats. But the wire of which pins are made acquires a perfectly cylindrical form by the simplest machinery. It is forcibly drawn through the circular holes of a steel plate; and the hole being smaller and smaller each time it is drawn through, it is at length reduced to the size required.

The head of a pin is a more difficult thing to make even than the body. It is formed of a small piece of wire twisted round so as to fit upon the other wire. It is said that by a machine fifty thousand heads can be made in an hour. We should think that a man would be very skilful to make fifty in an hour by hand, in the roughest manner ; if so, the machine does the work of a thousand men. The machine, however, does not do all the work. The head is attached to the body of a pin by the fingers of a child, while another machine rivets it on. The operations of cutting and pointing the pins are also done by machinery ; and they are polished by a chemical process.

It is by these processes,—by these combinations of human labour with mechanical power,—that it occurs that fifty pins can be bought for one halfpenny, and that therefore four or five thousand pins may be consumed in a year by the most economical housewife, at a much less price than fifty pins of a rude make cost two or three centuries ago. A woman's allowance was formerly called her *pin-money*,—a proof that pins were a sufficiently dear article to make a large item in her expenses. If pins now were to cost a halfpenny a piece instead of being fifty for a halfpenny, the greater number of females would adopt other modes of fastening their dress, which would probably be less neat and convenient than pins. No such circumstance could happen while the machinery of pin-making was in use ; but if the machinery were suppressed, by any act of folly on the part of the pin-makers who work with the machinery, pins would go out of use, probably, altogether : the pin-makers would lose *all* their employment ; and all the women of the land would be deprived of one of the simplest, and yet most useful inventions connected with the dress of modern times.

Needles are not so cheap as pins, because the material of which they are made is more expensive, and the processes cannot be executed so fully by machinery. But without machinery, how could that most beautiful article, a *fine* needle, be sold at the rate of six for a penny ?

As in the case of pins, machinery is at work at the first formation of the material. Without the tilt-hammer, which beats out the bar of steel, first at the rate of ten strokes a minute, and lastly at that of five hundred, how could that bar be prepared for needle-making at anything like a reasonable price? In all the processes of needle-making, labour is saved by contrivance and machinery. What human touch would be exquisite enough to make the eye of the finest needle, through which the most delicate silk is with difficulty passed? Needles are made in such large quantities, that it is even important to save the time of the child who lays them all one way when they are completed. Mr. Babbage, who is equally distinguished for his profound science and his mechanical ingenuity, has described this process as an example of one of the simplest contrivances which can come under the denomination of a tool. " It is necessary to separate the needles into two parcels, in order that their points may be all in one direction. This is usually done by women and children. The needles are placed sideways in a heap, on a table, in front of each operator. From five to ten are rolled towards this person by the fore-finger of the left hand ; this separates them a very small space from each other, and each in its turn is pushed lengthways to the right or to the left, according as its eye is on the right or the left hand. This is the usual process, and in it every needle passes individually under the finger of the operator. A small alteration expedites the process considerably ; the child puts on the fore-finger of its right hand a small cloth cap or finger-stall, and rolling from the heap from six to twelve needles, it keeps them down by the fore-finger of the left hand ; whilst it presses the fore-finger of the right hand gently against the ends of the needles, those which have their points towards the right hand stick into the finger-stall ; and the child, removing the finger of the left hand, allows the needles sticking into the cloth to be slightly raised, and then pushes them towards the left side. Those needles which had their eyes on the right hand do not stick into the finger cover, and are pushed to the heap on the right side previous to the repetition of this process. By means of this simple contrivance, each movement of the finger, from one side to the other, carries five or six needles to their proper heap ; whereas, in the former method, frequently only one was moved, and rarely more than two or three were transported at one movement to their place."

We have selected this description of a particular process in needle-making, to show that great saving of labour may be effected by what is not popularly called machinery. In modern times, wherever work is carried on upon a large scale, the division of labour is applied ; by which one man attending to one thing learns to perform that one thing more perfectly than if he had attended to many things. He thus saves a considerable portion of the whole amount of labour. Every skilful workman has individually some mode of working peculiar to himself, by which he lessens his labour. An expert blacksmith, for instance, will not strike one more blow upon the anvil than is necessary to produce the effect he desires. A compositor, or printer who arranges the types, is a swift workman when he makes no unnecessary movement of his arm or fingers in lifting a single type into what is called his composing-stick, where the types are arranged in lines. There is a very simple contrivance to lessen the labour of the compositor, by preventing him putting the type into his composing stick the wrong side outwards. It is a nick or two nicks, on the side of the type which corresponds with the lower side of the face of the letter. By this nick or nicks he is enabled to see by one glance of his eye on which side the letter is first to be grasped, and then to be arranged. If the nick were not there he would have to look at the face of every letter before he could properly place it. Now if the printers, as a body, were to resolve to perform their work in a difficult instead of an easy way,—if they were to resolve, that the labour employed in printing were desirable to be doubled,—they might effect their unwise resolution by the simplest proceeding in the world. They might refuse to work upon types which had any

nicks. In that case two compositors would certainly be
required to do the work of one ; and the price of printing would consequently be greatly raised, if the compositors were paid at the present rate for their time.
But would the compositors, who thus rejected one of the
most obvious natural aids to their peculiar labour, be
benefited by this course ? No. For the price of books
would rise, in the same proportion that the labour required to produce them was doubled in its quantity, by
being lessened one half in its efficiency. And the price
of books rising, and that rise lessening purchasers, thousands of families would be deprived of a livelihood ;—
not only those of compositors, but those of paper-makers,
type-founders, and many other trades connected with
books.

If, however, machines are bad things because they save
human labour, so are all these arrangements. A manufactory, where one man does one part, and another
another, is a human machine, in which one person is a
wheel, another a strap, a third a lever, and so forth. If
one person were to perform every operation in making a
pin or needle, he could not make ten in a day—probably
not one. It is said that among the early settlers of North
America, there was once a whole village in which there
was but one needle. If prejudices against machinery
should extend to everything which economizes human
labour, and be powerful to derange that economy, an
English village might be found in the same predicament.

Contrivances such as that of the needle-sorter's sheath
are constantly occurring in manufactures. The tags of
laces, which are made of thin tin, are now bent into their
requisite form by the same movement of the arm that cuts
them. A piece of steel, adapted to the side of the shears,
gives them at once their proper shape. All such ingenious applications of scientific principles lessen the price
of a commodity. If the small shot which is used by
sportsmen were each cast in a mould, the price would be
enormous ; but by pouring the melted lead, of which the
shot is made, through a sort of cullender, placed at the
top of a tower, high enough for the lead to cool in its

passage through the air, before it reaches the ground, the shot is formed in a spherical or round shape, by the mere act of passing through the atmosphere. Some of the shots thus formed are not perfectly spherical—they are pear-shaped. If the selection of the perfect from the imperfect shots were made by the eye, or the touch, the process would be very tedious and insufficient, and the price of the article much increased. The simplest contrivance in the world divides the bad from the good. The shots are poured down an inclined plane, and, without any trouble of selection, the spherical ones run straight to the bottom, while the pear-shaped ones tumble off on one side or the other of the plane.

A vast number of people at Birmingham are employed in the manufacture of buttons; and a great variety of hands are employed in the manufacture of a single button, such as piercers, cutters, stampers, gilders, and burnishers. Many of the operations in button-making are performed by machinery. The shanks are made by a little machine worked by a steam-engine, at the rate of fourscore a minute. But do these engines throw the button-makers out of employ? On the contrary, the cheapness by which the shank is made by the machine, instead of being expensively made by the slow labour of the hand, allows all sorts of hand-workmen to complete the rest of the button, whether in metal or glass; and thus Birmingham buttons are sold all over the world.

The manufacture of trinkets, and small articles of taste, at a cheap price by machinery, creates a demand for such articles, that could never have existed at all, if they were made by hand; and therefore creates employment, which could never otherwise have existed, for very numerous workmen.

In 1824, Mr. Osler, an intelligent manufacturer at Birmingham, showed a Committee of the House of Commons an imitation, in coloured glass, of an engraving on stone. "Those impressions," said he, "could not be given to real stones, but at the expense of from a guinea to thirty shillings each, if executed by a tolerable artist. We produce them for three halfpence, and the mounting of them furnishes employ to a very large number of hands indeed."

The application of machinery, or of peculiar scientific modes of working, to such apparently trifling articles as pins, needles, buttons, and trinkets, may appear of little importance. But let it be remembered, that the manufacture of such articles furnishes employment to many thousands of our fellow-countrymen ; and, enabling us to supply other nations with these products, affords us the means of receiving articles of more intrinsic value in exchange. In 1842, our exports of hardware, cutlery, and brass goods amounted to upwards of three millions sterling. No article of ready attainment, and therefore of general consumption, whether it be a labourer's spade, or a child's marble, is unimportant in a commercial point of view. The wooden figures of horses and sheep that may be bought for twopence in the toy-shops, furnish employment to cut them, during the long winter nights, to a large portion of the peasantry of the Tyrol (an extensive district on the boundaries of Austria). The insignificant article of the eyes of children's dolls, alone, produce, in their manufacture, a circulation of several thousand pounds. Mr. Osler, whose words we have just quoted, addressing a Committee of the House of Commons, upon the subject of his beads and trinkets, said,—" Eighteen years ago, on my first journey to London, a respectable looking man in the city asked me if I could supply him with dolls' eyes ; and I was foolish enough to feel half offended. I thought it derogatory to my new dignity as a manufacturer to make dolls' eyes. He took me into a room quite as wide and perhaps twice the length of this (one of the large rooms for Committees in the House of Commons), and we had just room to walk between stacks, from the floor to the ceiling, of parts of dolls. He said, ' These are only the legs and arms—the trunks are below.' But I saw enough to convince me that he wanted a great many eyes ; and as the article appeared quite in my own line of business, I said I would take an order by way of experiment ; and he showed me several specimens. I copied the order. He ordered various quanti-

ties and of various sizes and qualities.   On returning to
the Tavistock Hotel, I found that the order amounted to
upwards of five hundred pounds."

Mr. Osler tells this story to show the importance of
trifles.   The making of dolls' eyes afforded subsistence
to many ingenious workmen in glass toys; and in the
same way the most minute and apparently insignificant
article of general use, when rendered cheap by chemical
science or machinery, produces a return of many thousand pounds, and sets in motion labour and labourers.
Without the science and the machinery, which render
the article cheap, the labourers would have had *no* employ, for the article would not have been consumed.
What a pretty article is a common tobacco-pipe, of which
millions are used !   It is made cheap and beautiful in a
mould—a machine for copying pipes.   If the pipe were
made without the mould, and other contrivances, it would
cost at least a shilling instead of a halfpenny :—the
tobacco-smoker would go without his pipe, and the pipe-
maker without his employment.

CHAPTER XXIII.

WE exhibited in the last chapter a few examples, such as the sheath of the needle-sorter, and the nicks in the types of the compositor, of contrivances to economize labour. Such contrivances are not machinery; but they answer one of the great purposes of machinery,— that of saving time; and in the same manner they diminish the cost of production. The objection which some make to machinery, namely, that it diminishes the quantity of labour required, and therefore the number of labourers, applies also to these contrivances; and it applies, also, to the greater expertness of one workman as compared with the lesser expertness of another workman. There are boot-closers so skilful that they have reduced their arms to the precision of a machine. They can begin to close a boot with a thread a yard long in each hand, throw out each arm at once to the full extent of the thread, without making a second pull, and at every successive pull contract the arm so as to allow for the diminished length of the thread each time that it passes through the leather. There are not many workmen who can do this; but those whose sense of touch is delicate enough are not blamed by their fellow-workmen, for doing that by one movement of the arm which other men do by two movements.

Every one of us who thinks at all is constantly endeavouring to diminish his individual labour, by the use of some little contrivance which experience has suggested. Men who carry water in buckets, in places where water is scarce, put a circular piece of wood to float on the water, which prevents its spilling, and consequently lessens the labour. A boy who makes paper bags in a grocer's shop, so arranges them that he pastes the edges of twenty at a time, to diminish the labour. The porters of Amsterdam, who draw heavy goods upon a sort of sledge, every now and then throw a greased rope under the sledge, to diminish its friction, and therefore to lessen the labour of dragging it. Other porters, in the same city, have a little barrel containing water, attached to each side of the sledge, out of which the water slowly drips, like the water upon a stone-cutter's saw, to diminish the friction. The dippers of candles have made several improvements in their art within the last twenty years, for the purpose of diminishing labour. They used to hold the rods between their fingers, dipping three at a time; they next connected six or eight rods together by a piece of wood at each end, having holes to receive the rods; and they now suspend the rods so arranged upon a sort of balance, rising and falling with a pulley and a weight, so as to relieve the arms of the workman almost entirely, while the work is done more quickly and with more precision. Are there fewer candle-makers employed now than when they dipped only three rods with considerable fatigue, and no little pain as the candles grew heavy?

In the domestic arrangements of a well-regulated household, whether of a poor man or of a rich man, one of the chief cares is, to save labour. Every contrivance to save labour that ingenuity can suggest is eagerly adopted when a country becomes highly civilized. In former times, in our own country, when such contrivances were little known, and materials as well as time were consequently wasted in every direction, a great Baron was surrounded with a hundred menial servants; but he had certainly less real and useful labour performed for him, than a tradesman of the present day obtains from three servants. Are there fewer servants now employed than in those times of barbarous state? Certainly not. The middle classes amongst us can get a great deal done for them in the way of domestic service, at a small expense; because servants are assisted by manifold contrivances which do much of the work for them. The contrivances render the article of service cheaper; and therefore there are more servants. The work being done by fewer servants, in consequence of the contrivances, the servants themselves are better paid than if there was no cost saved by the contrivances.

The common jack by which meat is roasted is described by Mr. Babbage as "a contrivance to enable the cook in a few minutes to exert a force (in winding up the jack) which the machine retails out during the succeeding hour in turning the loaded spit, thus enabling her to bestow her undivided attention on her other duties." We have seen, twenty years ago, in farm-houses, a man employed to turn a spit with a handle; dogs have been used to run in a wheel for the same purpose, and hence a particular breed so used are called "turn-spits." When some ingenious servant girl discovered that if she put a skewer through the meat and hung it before the fire by a skein of worsted, it would turn with very little attention, she made an approach to the principle of the bottle-jack. All these contrivances diminish labour, and ensure regularity of movement;—and therefore they are valuable contrivances.

A bell which is pulled in one room and rings in another, and which therefore establishes a ready communication between the most distant parts of a house, is a contrivance to save labour. In a large family, the total want of bells would add a fourth at least to the labour of servants. Where three servants are kept now, four servants would be required to be kept then. Would the destruction of all the bells therefore add one fourth to the demand for servants? Certainly not. The funds employed in paying for service would not be increased a single farthing; and, therefore, by the destruction of bells, all the families of the kingdom would have some work left undone, to make up for the additional labour required through the want of this useful contrivance: or all the servants in the kingdom would be more hardly worked,—would have to work sixteen hours a day instead of twelve.

In some parts of India, the natives have a very rude contrivance to mark the progress of time. A thin metal cup, with a small hole in its bottom, is put to float in a vessel of water; and as the water rises through the hole the cup sinks in a given time—in 24 minutes. A servant is set to watch the sinking of the cup, and when this happens he strikes upon a bell. Half a century ago, almost every cottage in England had its hour-glass—an imperfect instrument for registering the progress of time, because it only indicated its course between hour and hour; and an instrument which required a very watchful attention, and some labour, to be of any use at all. The universal use of watches or clocks, in India, would wholly displace the labour of the servants who note the progress of time by the filling of the cup; and the same cause has displaced, amongst us, the equally unprofitable labour employed in turning the hour-glass, and watching its movement. Almost every house in England has now a clock or watch of some sort; and every house in India would have the same, if the natives were more enlightened, and were not engaged in so many modes of unprofitable labour to keep them poor. His profitable labour has given the English mechanic the means of getting a watch. Machinery, used in every possible way, has made this watch cheap. The labour formerly employed in turning the hour-glass, or in running to look at the church clock, is transferred to the making of watches. The user of the watch obtains an accurate register of time, which teaches him to know the value of that most precious possession, and to economize it; and the producers of the watch have abundant employment in the universal demand for this valuable machine.

A watch or clock is an instrument for assisting an operation of the mind. Without some instrument for registering time, the mind could very imperfectly attain the end which the watch attains, not requiring any mental labour. The observation of the progress of time, by the situation of the sun in the day, or of particular stars at night, is a labour requiring great attention, and various sorts of accurate knowledge. It is therefore never attempted, except when men have no machines for registering time. In the same manner the labours of the mind have been saved, in a thousand ways, by other contrivances of science.

The foot-rule of the carpenter not only gives him the standard of a foot measure, which he could not exactly ascertain by any experience, or any mental process ; but it is also a scale of the proportions of an inch, or several inches, to a foot, and of the parts of an inch to an inch. What a quantity of calculations, and of dividing by compasses, does this little instrument save the carpenter ; besides ensuring a much greater degree of accuracy in all his operations ! The common rules of arithmetic, which almost every boy in England now learns, are parts of a great invention for saving mental labour. The higher branches of mathematics, of which science arithmetic is a portion, are also inventions for saving labour, and for doing what could never be done without these inventions. There are instruments, and very curious ones, for lessening the labour of all arithmetical calculations ; and tables, that is, the results of certain calculations, which are of practical use, are constructed for the same purpose. When we buy a joint of meat, we often see the butcher turn to a little book, before he tells us how much a certain number of pounds and ounces amounts to, at a certain price per pound. This book is his ' Ready Reckoner,' and a very useful book it is to him ; for it enables him to dispatch his customers in half the time that he would otherwise require, and thus to save himself a great deal of labour, and a great deal of inaccuracy. The inventions for saving mental labour, in calculations of arithmetic, have been carried so far, that Mr. Babbage, a gentleman whose name we have twice before mentioned, has almost perfected a calculating machine, which not only does its work of calculation without the possibility of error, but absolutely arranges printing types of figures, in a frame, so that no error can be produced in copying the calculations, before they are printed. We mention this curious machine, to show how far science may go in diminishing mental labour, and ensuring accuracy.

To all who read this book it is no difficulty to count a hundred ; and most know the relation which a hundred bears to a thousand, and a thousand bears to a million. Most are able, also, to read off those numbers, or parts of those numbers, when they see them marked down in figures. There are many uncivilized people in the world who cannot count twenty. They have no idea whatever of numbers, beyond perhaps as far as the number of their fingers, or their fingers and their toes. How have we obtained this great superiority over these poor savages ? Because science has been at work, for many centuries, to diminish the amount of our mental labour, by teaching us the easiest modes of calculation. And how did we learn these modes ? We learnt them from our school-masters.

If any follow up the false reasoning which has led some to think that whatever diminishes labour diminishes the number of labourers, they might conclude, that, as there is less mental work to be done, because science has diminished the labour of that work, there would, therefore, be fewer mental workmen. Thank God, the greater facilities that have been given to the cultivation of the mind, the greater is the number of those who exert themselves in that cultivation. The effects of saving unprofitable labour are the same in all cases. The use of machinery in aid of *bodily* labour has set that bodily labour to a thousand new employments ; and has raised the character of the employments, by transferring the lowest of the drudgery to wheels and pistons. The use of science in the assistance of *mental* labour has conducted that labour to infinitely more numerous fields of exertion ; and has elevated all intellectual pursuits, by making their commoner processes the play of childhood, instead of the toil of manhood.

CHAPTER XXIV.

WE cannot doubt that any invention which gives assistance to the thinking powers of mankind, and, therefore, by dispensing with much mental drudgery, leads the mind forward to nobler exertions, is a benefit to all. It is not more than four hundred years ago, that the use of Arabic numerals, or figures, began to be generally known in this country. The first date in those numerals said to exist in England, is upon a brass plate in Ware church, 1454. The same date in Roman numerals, which were in use before the Arabic ones, would be expressed by eight letters, MCCCCLIV. The introduction of figures, therefore, was an immense saving of time in the commonest operations of arithmetic. How puzzled we should be, and what a quantity of labour we should lose, if we were compelled to reckon earnings and marketings by the long mode of notation, instead of the short one! This book is easily read, because it is written in words composed of twenty-four letters. In China, where there are no letters in use, every word in the language is expressed by a different character. Few people in China write or read; and those who do, acquire very little knowledge, except the mere knowledge of writing and reading. All the time of their learned men is occupied in acquiring the means of knowledge, and not knowledge itself; and the bulk of the people get very little knowledge at all. It would be just the same thing if there were no machines or engines for diminishing manual labour. Those who had any property would occupy all their time, and the time of their immediate dependents,

in raising food and making clothes for themselves, and
the rest of the people would go without any food or
clothes at all ; or rather, which comes to the same thing,
there would be *no* " rest of the people ;" the lord and
his vassals would have all the produce ;—there would be
half a mil. n of people in England instead of sixteen
millions.

When a b has got hold of what we call the rudi-
ments of learni. g, he has possessed himself of the most
useful tools and machines which exist in the world. He
has obtained the means of doing that with extreme ease,
which, without these tools, is done only with extreme
labour. He has earned the time which, if rightly employed, will elevate his mind, and therefore improve his
condition. Just so is it with all tools and machines for
diminishing bodily exertion. They give us the means
of doing that with comparative ease, which, without
them, can only be done with extreme drudgery. They
set at liberty a great quantity of mere animal power,
which, having then leisure to unite with mental power,
produces ingenious and skilful workmen in every trade.
But they do more than this. They diminish human
suffering—they improve the health—they increase the
term of life—they render all occupations less painful
and laborious ;—and, by doing all this, they elevate man
in the scale of existence.

The present Pasha, or chief ruler of Egypt, in one of
those fits of caprice which it is the nature of tyrants to
exhibit, ordered, a few years ago, that the male population of a district should be set to clear out one of the
ancient canals which was then filled up with mud. The
people had no tools, and the Pasha gave them no tools ;
but the work was required to be done. So to work the
poor wretches went, to the number of fifty thousand.
They had to plunge up to their necks in the filthiest
slime, and to bale it out with their hands, and their
hands alone. They were fed, it is true, during the operation ; but their food was of a quality proportioned to
the little *profitable* labour which they performed. They
were fed on horse-beans and water. In the course of

one year, more than thirty thousand of these unhappy people perished. If the tyrant, instead of giving labour to fifty thousand people, had possessed the means of setting up steam-engines to pump out the water, and scoop out the mud,—if he had even provided the pump, which is called Archimedes' screw, and was invented by that philosopher for the very purpose of draining land in Egypt,—if the people had even had scoops and shovels, instead of being degraded like beasts, to the employment of their unassisted hands,—the work might have been done at a fiftieth of the cost, even of the miserable pittance of horse-beans and water; and the money that was saved by the tools and machines, might have gone to furnish *profitable* labour to the thousands who perished amidst the misery and degradation of their *unprofitable* labour.

Some may say that this is a case which does not apply to us; because we are free men, and cannot be compelled to perish, up to our necks in mud, upon a pittance of horse-beans, doled out by a tyrant. Exactly so. But what has made us free? Knowledge. Knowledge,—which, in raising the moral and intellectual character of every Englishman, has raised up barriers to oppression which no power can ever break down. Knowledge,—which has set ingenious men thinking in every way how to increase the profitable labour of the nation, and therefore to increase the comforts of every man in the nation. Is it for the working men of this country, or for any other class of men, to say that knowledge shall stop at a certain point, and shall go no farther? Is it for them to say, that although they are willing to retain the infinite blessings which knowledge has bestowed upon them—the improved food, the abundant fuel and water, the cheap clothing, the convenient houses, the drainage and ventilation which make houses healthful, the preservation of life by medical science, and the profit and comfort of books—that we are to rest satisfied with what we have got: or rather, if the haters of machinery are to be heard, that we are to go back to what we were five hundred years ago? Depend upon

it, if we once begin to march backwards, however slow
may be the first steps, the retreat towards ignorance,
instead of the advance towards knowledge, will soon
become pretty quick ; till at last there would be one
mad rush from civilization to *un*civilization. Then comes
the labour of the despot, who has been comparatively
idle while knowledge was labouring. There is no halt-
ing-place *then ;* and the mud and horse-beans of the
Pasha of Egypt will be the natural end, and the fit reward, of such monstrous folly and wickedness.

No one, we suppose, desires to be sick instead of in
health, to live a short life instead of a long one. The
people of England have gone on increasing very rapidly
during the last fifty years ; and the average length of
life has also gone on increasing in the same remarkable
manner. Men who have attended to subjects of political
economy have always been desirous to procure accurate
returns of the average duration of life at particular
places, and they could pretty well estimate the condition
of the people from these returns. Savages, it is well
known, are not long livers ; that is, although there may
be a few old people, the majority of savages die very
young. Why is this? Many of the savage nations
that we know have much finer climates than our own ;
but then, on the other hand, they sustain privations
which the poorest man amongst us never feels. Their
supply of food is uncertain, they want clothing, they are
badly sheltered from the weather, or not sheltered at
all, they undergo very severe labour when they are
labouring. From all these causes savages die young. Is
it not reasonable, therefore, to infer that if in any particular country the average duration of life goes on increasing ; that is, if fewer people, in a given number and
a given time, die now than formerly, the condition of
that people is improved ; that they have more of the
necessaries and comforts of life, and labour less severely
to procure them? Now let us see how the people of
England stand in this respect. The average mortality
in a year about a century ago was reckoned to be one in
thirty, and now it is one in forty-six. This result is,

doubtless, produced in some degree by improvement in the science of medicine, and particularly by the use of inoculation for the small-pox, and vaccination. But making every allowance for these benefits, the fact furnishes the most undeniable truth, that the people of England are much better fed, clothed, and lodged than they were a century ago, and that the labour which they perform is far less severe.

The effect of continued violent bodily exertion upon the duration of life might be illustrated by many instances; we shall mention one. The late Mr. Edgeworth, in his Memoirs, repeatedly speaks of a boatman whom he knew at Lyons, as an old man. " His hair," says Mr. Edgeworth, " was grey, his face wrinkled, his back bent, and all his limbs and features had the appearance of those of a man of sixty; yet his real age was but twenty-seven years. He told me that he was the oldest boatman on the Rhone, that his younger brothers had been worn out before they were twenty-five years old ; such were the effects of the hardships to which they were subject from the nature of their employment." That employment was, by intense bodily exertion, and with the daily chance of being upset, to pull a boat across one of the most rapid rivers in the world,—

"The swift and arrowy Rhone,"

as one of our poets calls it. How much happier would these boatmen have been during their lives, and how much longer would they have lived, could their labour have been relieved by some mechanical contrivance! and without doubt, the same contrivance would have doubled the number of the boatmen, by causing the passage to be more used. As it was, they were few in number, they lived only a few years, and the only gratification of those few years was an inordinate stimulus of brandy. This is the case in all trades where immense efforts of bodily power are required. The exertion itself wears out the people, and the dram, which gives a momentary impulse to the exertion, wears them out still more. The coal-heavers of London, healthy as they look, are but a short-lived people. The heavy loads which they carry, and the quantity of liquor which they drink, both together make sad havoc with them.

Violent bodily labour, in which the muscular power of the body is unequally applied, generally produces some peculiar disease. Nearly all the pressmen who were accustomed to print newspapers of a large size, by hand, were ruptured. The printing-machine now does the same description of work.

What is the effect upon the condition of pressmen generally by the introduction of the printing-machine to do the heaviest labour of printing? That the trade of a pressman is daily becoming one more of *skill* than of *drudgery*. At the same time that the printing-machine was invented, one of the principles of that machine, that of inking the types with a roller instead of two large cushions, called balls, was introduced into hand-printing. The pressmen were delighted with this improvement. " Ay," said they, " this saves our labour ; we are relieved from the hard work of distributing the ink upon the balls." What the roller did for the individual pressman, the machine, which can only be beneficially applied to rapid and to very heavy printing, does for the great body of pressmen. It removes a certain portion of the drudgery, which degraded the occupation, and rendered it painful and injurious to health. We have seen two pressmen working a daily paper against time ; it was always necessary, before the introduction of the machines, to put an immense quantity of bodily energy into the labour of working a newspaper, that it might be published at the proper hour. Time, in this case, was driving the pressmen as fast as the rapid stream drove the boatmen of the Rhone ; and the speed with which they worked was killing them as quickly.

If artisans, who have generally the means of acquiring knowledge, were to think as they ought to do upon the benefits to their own particular trade of machines for saving labour, we should never hear of combinations against such machines. A reflecting being feels it a degradation to be employed in *unprofitable* labour. Some parishes, we understand, formerly set their paupers to turn a grindstone, upon which nothing is ground; and, to their honour be it spoken, the poor people, in many cases, would rather starve than submit to this ignoble occupation.   Even the unhappy persons at treadmills feel additionally degraded when they turn the wheel without an object; they call it " grinding the wind." Why are these people degraded by such occupations ? Why do they consider their labour ignominious ?   Because their labour has no results.   Is it not equally ignominious when men would resolve, by suppressing machinery, to do that with a great deal of labour, which would otherwise be done by a very little labour ? to bind themselves to the wheel, when the wheel would do the work without them ?—to labour, in fact, without results from their labour.

CHAPTER XXV.

Working men, and other men who ought to know better,
have attempted to draw distinctions between old machines
and new machines. As it is, the inventors of machines
generally go before their age ; and thus too many of
them have either starved or struggled for years with
want, because their own generation was not wise enough
to value the blessings which science and skill had provided for it. But if the ordinary difficulties of estab-
lishing a new invention, however valuable it might prove,
were to be increased by the folly which should say, we
will have no new machines at all, or at any rate, a machine shall become old before we will use it, there would
be an end to invention altogether.

Before the invention of the first stocking-machine, in
the year 1589, by William Lea, a clergyman, none but
the very rich wore stockings, and many of the most
wealthy went without stockings at all, their hose being
sewn together by the tailor, or their legs being covered
with bandages of cloth. William Lea made a pair of
stockings, by the frame, in the presence of King James
I. ; but such was the prejudice of those times, that he
could get no encouragement for his invention. His invention was discountenanced, upon the plea that it would
deprive the industrious poor of their subsistence. He
went to France, where he met with no better success,
and died at last of a broken heart. The great then *could*
discountenance an invention, because its application was
limited to themselves. *They* only wore stockings : the
poor who made them had none to wear. Stockings were

not cheap enough for the poor to wear, and therefore they went without. Of the millions of people now in this country, how few are without stockings! What a miserable exception to the comfort of the rest of the people does it appear when we see a beggar in the streets without stockings! We consider such a person to be in the lowest stages of want and suffering. Two centuries ago, not one person in a thousand wore stockings ;—one century ago, not one person in five hundred wore them ;—now, not one person in a thousand is without them. Who made this great change in the condition of the people of England, and, indeed, of the people of almost all civilized countries? William Lea—who died at Paris of a broken heart. And why did he die of grief and penury? Because the people of his own days were too ignorant to accept the blessings he had prepared for them.

We ask with confidence, had the terror of the stocking-frame any real foundation? Were any people thrown out of employment by the stocking-frame?

> " The knitters in the sun,
> And the free maids who weave their thread with bones,"

as Shakspere describes the country lasses of his day, had to *change* their employment; but there was far more employment for the makers of stockings, for then every one began to wear stockings.

But suppose that the ignorance and prejudice which prevailed at the time of James I. upon the subject of machinery, had continued to the present day; and that not only the first stocking-frame of William Lea had never been used, but that all machines employed in the manufacture of hosiery had never been thought of; and they could not have been thought of, if the first machines had been put down. The greater number of us, in that case, would have been without stockings.

But there would have been a greater evil than even this. We might all have found substitutes for stockings, or have gone without them. But the progress of ingenuity would have been stopped. The inventive principle would have been destroyed. Society must either go forward or back-ward. There can be no halting place for any long period. If we had gone backward, we should not only have lost stockings, but all the comforts—all the decen-cies—all the elegancies—and, worse than these losses, all the knowledge—which distinguish the civilised from the uncivilised state. The same thing would happen *now*, if the principle were admitted that new inventions, and new machines, are *evils*, and not benefits. The stocking-frame was once a *new* machine, and, therefore, the court discountenanced it. There are people at the present day as ignorant as the court was then, who would discounte-nance new machines, that, like the stocking-machine, will some day be *old*. An engineer, who has contributed largely to benefit society by his inventions, told the writer of this book, in the latter part of 1830, that he had com-pleted several machines which he considered of general utility, but which he dared not bring forward in the state of the popular feeling which then prevailed. If this feeling were to prevail and to extend ;—if the brute force which seeks to destroy machinery were not to be put down by the power of the laws, and if the unwise preju-dice which desires to repress it could not be conquered by the power of reason—the glory and prosperity of this country would be gone for ever. We should have reached the end of our career of improvement.—We should begin a backward race ; and it would remain for the inquiring savages of such countries as New Zealand and Otaheite to march forward. The night of the dark ages would return to Europe.

Those who have not taken the trouble to witness, or to inquire into, the processes by which they are sur-rounded with the conveniences and comforts of civilised life, can have no idea of the vast variety of ways in which invention is at work to lessen the cost of produc-tion. The people of India, who spin their cotton wholly by hand, and weave their cloth in a rude loom, would doubtless be astonished when they first saw the effects of machinery, in the calico which is returned to their own shores, made from the material brought from their own shores, cheaper than they themselves could make it. But their indolent habits would not permit them to inquire how machinery produced this wonder. There are many amongst us who only know that the wool grows upon the sheep's back, and that it is converted into a coat by labour and machinery. They do not estimate the prodigious power of thought—the patient labour—the unceasing watchfulness—the frequent disappointment—the uncertain profit—which many have had to encounter in bringing this machinery to perfection. How few, even of the best informed, know that in the cotton manufacture, which from its immense amount possesses the means of rewarding the smallest improvement, invention has been at work, and most successfully, to make machines, that make machines, that make the cotton thread! There is a part of the machinery used in cotton-spinning called a reed. It consists of a number of pieces of wire, set side by side in a frame, resembling, as far as such things admit of comparison, a comb with two backs. These reeds are of various lengths and degrees of fineness; but they all consist of cross pieces of wire, fastened at regular intervals between longitudinal pieces of split cané, into which they are tied with waxed thread. A machine now does the work of reed-making. The materials enter the machine in the shape of two or three yards of cane, and many yards of wire and thread; and the machine cuts the wire, places each small piece with unfailing regularity between the canes, twists the thread round the cane with a knot that cannot slip, every time a piece of wire is put in, and does several yards of this extraordinary work in less time than we have taken to write the description. There is another machine for making a part of the machine for cotton spinning, even more wonderful. The cotton wool is combed by circular cards of every degree of fineness; and the card-making machine, receiving only a supply of leather and wire, does its own work without the aid of hands. It punches the leather—cuts the wire—passes it through the leather—clinches it behind,—and gives it the proper form of the tooth in front,—producing a complete card of several feet in circumference in a wonderfully short time. All men feel the benefit of such inventions, because they lessen the cost of production. The necessity for them always precedes their use. There were not reed-makers and card-makers enough in England to supply the demands of the cotton machinery ; so invention went to work to see how machines could make machines; and the consequent diminished cost of machinery has diminished the price of clothing.

Machinery enters into competition with human labour ; and therefore there are some people who say—let us tax machinery to support the labour which it supersedes. The real meaning of this is—let us tax machinery, to prevent cheapness of production, to discourage invention, and to interfere with a change from one mode of labour to another mode. There are temporary inconveniences doubtless, in machinery, and there are partial evils, which sometimes fall severely upon classes and individuals. But it appears to us that any proposed remedy for a temporary and partial evil, which has a tendency to arrest the course of improvement, is a little like the ancient wisdom of the Dutch market-woman, who, when the one pannier of her ass is too heavily laden with cabbages, puts a stone into the other pannier, to make matters equal.

CHAPTER XXVI.

Lord Bacon, the great master of practical wisdom, has said that " the effort to extend the dominion of man over nature is the most healthy and most noble of all ambitions."—" The empire of man," he adds, " over material things has for its only foundation the sciences and the arts." A great deal of the knowledge which constitutes this dominion has been the property of society, handed down from the earliest ages. No one can tell, for instance, how the art of leavening bread was introduced amongst mankind ; and yet this process, now so familiar to all, contributes as much as, if not more than, any other art to the wholesome and agreeable preparation of our food. Leavening bread is a branch of chemistry ; and, like that process, many other processes of chemistry have been the common property of civilized man from time immemorial. Within a few centuries, however, science has applied its discoveries to the perfection of the arts ; and in proportion as capital has been at hand to encourage science, has the progress of the application been certain and rapid. The old Alchemists, or hunters after the philosopher's stone, sought to create capital by their discoveries. They could not make gold, but they discovered certain principles which have done as much for the creation of utility in a few hundred years as the rude manual labour of all mankind during the same period. Let it not be supposed that we wish to depreciate manual labour. We only wish to show that labour is incomparably more prolific when directed by science. Mahomet Bey, the ruler of Tunis, was dethroned by his subjects. He was a clever man, and had the reputation of possessing the philosopher's stone, or the art of turning common metals into gold. The Dey of Algiers restored him to his throne upon condition that the secret should be communicated to him. Mahomet, with great pomp and solemnity, sent the Dey of Algiers a plough. This was so far well. He intimated that to compel production by labour is to make a nation rich. But had he been able to transmit some of the science which now controls and guides the operations of the plough—the chemical knowledge which teaches the proper application of manures to soils—the rotation of crops introduced by the turnip-husbandry, which renders it unnecessary that the ground should ever be idle, — he would have gone farther towards communicating the real philosopher's stone.

The indirect influence, too, of a general advance in knowledge upon the particular advance of any branch of labour, is undeniable ;—for the inquiring spirit of an age spreads itself on all sides, and improvement is carried into the most obscure recesses, the darkest chinks and corners of a nation. It has been wisely and beautifully said, "We cannot reasonably expect that a piece of woollen cloth will be wrought to perfection in a nation which is ignorant of astronomy, or where ethics are neglected." * The positive influence of science in the direction of labour is chiefly exhibited in the operations of mechanics and chemistry applied to the arts, in the shape of machines for saving materials and labour, and of processes for attaining the same economy. We have described at great length the effects of these manifold inventions in the improvement of the condition both of producers and consumers. But there are many particulars in which knowledge has laboured, and is still labouring, for the advance of the physical and moral condition of us all, which may have escaped attention; because these labours operate remotely and indirectly, though

Hume's Essays.

not without the highest ultimate certainty and efficiency, in aiding the great business of production. These are the influences of science upon labour, not so direct as the mechanical skill which has contrived the steam-engine, or so indirect as the operation of ethics upon the manufacture of a piece of woollen cloth; but which confer a certain, and in some instances enormous benefit upon production, by the operation of causes which, upon a superficial view, appear to be only matters of laborious but unprofitable speculation. If we succeed in pointing out the extent and importance of those aids which production derives from the labours of men who have not been ordinarily classed amongst "working men," but who have been truly the hardest and most profitable workers which society has ever possessed, we shall show what an intimate union subsists amongst those classes of society who appear the most separated; and that these men really labour with all most effectually in the advancement of the great interests of mankind.

When Hume thought that a nation would be behind in the manufacture of cloth that had not studied astronomy, he perhaps did not mean to go the length of saying, that the study of astronomy has a real influence in making cloth cheaper, in lessening the cost of production, and in therefore increasing the number of consumers. But look at the direct influence of astronomy upon navigation. A seaman, by the guidance of principles laid down by the great minds that have directed their mathematical powers to the study of astronomy—such minds as those of Newton and La Place—measures the moon's apparent distance from a particular star. He turns to a page in the 'Nautical Almanac,' and, by a calculation, directed principally by this table, can determine whereabout he is upon the broad ocean, although he may not have seen land for three months. Sir John Herschel, a mathematician of our own times, who has united to the greatest scientific reputation the rare desire to make the vast possessions of the world of science accessible to all, has given, in his 'Discourse on the Study of Natural Philosophy,' an instance of the accuracy of such lunar

K

observations, in an account of a voyage of eight thousand miles, by Captain Basil Hall, who, without a single landmark during eighty-nine days, ran his ship into the harbour of Rio as accurately, and with as little deviation, as a coachman drives his stage into an inn-yard. But what has this, it may be said, to do with the price of clothing ?—Exactly this : part of the price arises from the cost of transport. If there were no " lunar distances " in the ' Nautical Almanac,' the voyage from New York to Liverpool might require three months instead of three weeks. But go a step farther back in the influence of science upon navigation. There was a time when ships could hardly venture to leave the shore. In the days of our Anglo-Saxon ancestors, a merchant who went three times over sea with his own craft might become a thegn, or nobleman. The honour thus bestowed was doubtless to encourage persons of property to encounter an enterprise then held of great difficulty. Long after this early period of England's navigation, voyages across the Atlantic could never have been attempted. That was before the invention of the mariner's compass ; but even after that invention, when astronomy was not scientifically applied to navigation, long voyages were considered in the highest degree dangerous. The crews both of Vasco de Gama, who discovered the passage to India, and of Columbus, principally consisted of criminals, who were pardoned on condition of undertaking a service of such peril. The discovery of magnetism, however, changed the whole principle of navigation, and raised seamanship to a science. If the mariner's compass had not been invented, America could never have been discovered ; and if America, and the passage to India by the Cape of Good Hope, had never been discovered, cotton would never have been brought to England ; and if cotton had never been brought to England, we should have been as badly off for clothing as the people of the middle ages, and the million of working men and women, manufacturers of cotton, would have been without employment.

Astronomy, therefore, and navigation, both sciences the results of long ages of patient inquiry, have opened a

communication between the uttermost ends of the earth ;
and therefore have had a slow, but certain effect upon the
production of wealth, and the consequent diffusion of all
the necessaries, comforts, and conveniences of civilized
life.   The connexion between manufactures and science,
practical commerce and abstract speculation, is so intimate that it might be traced in a thousand striking instances.   Columbus, the discoverer of America, satisfied
his mind that the earth was round ; and when he had got
this abstract idea firmly in his head, he next became
satisfied that he should find a new continent by sailing
in a westerly course.   The abstract notion which filled
the mind of Columbus that the earth was a sphere, ultimately changed the condition of every living being in the
Old World that then existed, or has since existed.   In
the year 1488, the first geographical maps and charts
that had been seen in England were brought hither by
the brother of the great Christopher Columbus.   If these
maps had not been constructed by the unceasing labours
of men in their closets, Columbus would never have
thought of discovering " the unknown land " which
occupied his whole soul.   If the scanty knowledge of
geography which existed in the time of Columbus had
not received immense additions from the subsequent
labours of other students of geography, England would
not have twenty thousand merchant ships ready to trade
wherever men have anything to exchange,—that is,
wherever men are enabled to give of their abundance
for our abundance, each being immensely benefited by
the intercourse.   A map now appears a common thing,
but it is impossible to overrate the extent of the accumulated observations that go to make up a map.   An almanac seems a common thing, but it is impossible to
overrate the prodigious accumulations of science that go
to make up an almanac.   With these accumulations, it is
now no very difficult matter to construct a map or an
almanac.   But if society could be deprived of the accumulations, and we had to re-create and remodel everything for the formation of our map and our almanac, it
would perhaps require many centuries before these

accumulations could be built up again; and all the arts of life would go backward, for want of the guidance of the principles of which the map and the almanac are the interpreters for popular use.

Science, we see, connects distant regions, and renders the world one great commercial market. Science is, therefore, a chief instrument in the production of commercial wealth. But we have a world beneath our feet which science has only just now begun to explore. We want fuel and metallic ore to be raised from the bowels of the earth; and, till within a very few years, we used to dig at random when we desired to dig a mine, or confided the outlay of thousands of pounds to be used in digging, to some quack whose pretensions to knowledge were even more deceptive than a reliance upon chance. The science of geology, almost within the last quarter of a century, has been able, upon certain principles, to determine where coal especially can be found, by knowing in what strata of earth coal is formed; and thus the expense of digging through earth to search for coal, when science would at once pronounce that no coal was there, has been altogether withdrawn from the amount of capital to be expended in the raising of coal. That this saving has not been small, we may know from the fact, that eighty thousand pounds were expended fruitlessly in digging for coal at Bexhill, in Sussex, not many years ago, which expense geology would have instantly prevented; and have thus accumulated capital, and given a profitable stimulus to labour, by saving their waste.

Whatever diminishes the risk to life or health, in any mechanical operation, or any exertion of bodily labour, lessens the cost of production, by diminishing the premium which is charged by the producers to cover the risk. The safety-lamp of Sir Humphry Davy, by diminishing the waste of human life employed in raising coals, diminished the price of coals. The magnetic mask, which prevents iron-filings escaping down the throats of grinders and polishers, and thus prevents the consumption of the lungs, to which these trades are peculiarly obnoxious, would diminish the price of steel goods, if the workmen did not prefer receiving the premium in the shape of higher wages, to the health and long life which they would get, without the premium, by the use of the mask. This is not wisdom on the part of the workmen. But whether they are wise or not, the natural and inevitable influence of the discovery, sooner or later, to lessen the cost of production in that trade, by lessening the risk of the labourers, must be established. The lightning conductor of Franklin, which is used very generally on the Continent, and almost universally in shipping, diminishes the risk of property, in the same way that the safety-lamp diminishes the risk of life; and, by this diminution, the rate of insurance is lessened, and the cost of production therefore lessened. Lightning is one of the destructive forces of Nature, in particular cases, which science knows how to control. A few years ago, all the timber in the Hartz Forest in Germany was destroyed by a species of beetle, which, gnawing completely round the bark, prevented the sap from rising. This destructive animal made its appearance in England; and science very soon discovered the cause of the evil, and provided for its removal. If there had been no knowledge of natural history here, not a tree would have been left in our woods;—and what then would have been the cost of timber?

It has been said by an American writer, who has published several treatises well calculated to give the workman an elevated idea of his rights and duties, that the " man who will go into a cotton-mill,—who will observe the parts of the machinery, and the various processes of the fabric, till he reaches the hydraulic press, with which it is made into a bale, and the canal or railroad by which it is sent to market, may find every branch of trade, and every department of science, literally crossed, intertwined, interwoven with every other, like the woof and the warp of the article manufactured."* This crossing and intertwining of the abstract and practical sciences,

* ' Everett's Working Man's Party.' Printed in the American Library of Useful Knowledge, 1831.

the mechanic skill and the manual labour, which are so striking in the manufacture of a piece of calico, prevail throughout every department of industry in a highly civilized community. Every one who labours at all profitably, labours for the production of utility, and sets in motion the labour of others. Look at the labour of the medical profession. In the fourteenth century, John de Gaddesden treated a son of Edward II. for the small-pox, by wrapping him up in scarlet cloth, and hanging scarlet curtains round his bed; and, as a remedy for epilepsy, the same physician carried his patients to church to hear mass! The medical art was so little understood in those days, that the professors of medicine had made no impression upon the understanding of the people; and they consequently trusted not to medicine, but to vain charms, which superstitions the ignorance of the practitioners themselves kept alive. Francis I., King of France, having a persuasion that, because the Jews were the most skilful physicians of that day, the virtue was in the Jew, and not in the science which he professed, sent to Charles V. of Spain for a Jewish phy-sician; but finding that the man who arrived had been converted to Christianity, he refused to employ him, thinking the virtue of healing had therefore departed from him. A statute of Henry VIII. says, "For as much as the science and cunning of physic and surgery is daily within this realm exercised by a great multitude of ignorant persons, of whom the greater part have no insight in the same, nor in any other kind of learning: some also con no letters on the book, so far forth, that common artificers, as smiths, and weavers, and women, boldly and accustomably, take upon them great cures, in which they partly use sorcery and witchcraft; partly apply such medicines to the disease as be very noxious, and nothing meet, to the high displeasure of God, great infamy to the faculty, and the grievous damage and destruction of diverse of the king's people." When such ignorance prevailed, diseases of the slightest kind must have been very often fatal; and the power of all men to labour profitably must have been greatly diminished by the ravages of sickness. These ravages are now checked by medical science and medical labour. In some parts of France, even now it is said, the sick cattle are taken by the country people outside the churches to hear mass. If there were a skilful cattle-doctor in those villages, the loss of stock would be doubtless lessened, and the people would be all the richer for the cattle-doctor's art, instead of degrading religion by a homage akin to the worst idolatry.

But the principle of connecting every exertion of the mind with profitable labour goes farther even than the applications of science, direct or indirect, to the mechanical arts. The sciences and arts cannot be carried forward except in a country where the laws of God are respected, where justice is upheld, where intellect generally is cultivated, and taste is diffused. The religious and moral teacher, therefore, who lifts the mind to a contemplation of the duties of man, as they are founded upon a belief in the Providence of an all-wise and all-powerful Creator, is a profitable labourer. The instructor of the young, who dedicates his time to advancing the formation of right principles, and the acquirement of sound knowledge, by his pupils, is a profitable labourer. The writer who applies his understanding to the discovery and dissemination of moral and political truth, is a profitable labourer. The interpreter and administrator of the laws, who upholds the reign of order and security, defending the innocent, punishing the guilty, and vindicating the rights of all from outrage and oppression, is a profitable labourer. These labourers, it may be said, are still direct producers of utility, but that those who address themselves to the imagination— the poets, the novelists, the painters, and the musicians, —in every polished society, are unprofitable labourers. One word is sufficient for an answer. These men advance the general intellect of a country, and they therefore indirectly advance the production of articles of necessity. We have already shown how the study of the higher mathematics, upon which astronomy is founded, has an influence upon the production of a piece of wool-

len cloth ; and we beg our readers to bear this connexion
in mind, when they hear it said, as they sometimes may,
that an abstract student, or an elegant writer, is not a
producer—is, in fact, an idler.   The most illustrious
writers of every country, the great poets,

> " High actions and high passions best describing,"

have, next to the inspirations of religion, lifted mankind,
more than any other class of intellectual workmen, to
their noblest pursuits of knowledge and virtue.   Even
the less dignified labourers in the same field,—those who
especially devote themselves to give pleasure and amusement,—call into action some of the highest and purest
sources of enjoyment.   They lead the mind to seek its
recreations in more ennobling pursuits than those of sen-
suality ; their arts connect themselves by a thousand
associations with all that is beautiful in the natural
world ; they are as useful for the promotion of pure and
innocent delight, as the flowers that gladden us by their
beauty and fragrance by the side of the corn that nou-
rishes us.   An entire community of poets and artists
would be as unprofitable as if an entire country were
dedicated to the cultivation of violets and roses ; but the
poets and the artists may, as the roses and the violets,
furnish the graces and ornaments of life, without injury,
and indeed with positive benefit, to the classes who more
especially dedicate themselves to what is somewhat exclusively called the production of utility.   The right
direction of the talents which are dedicated to art and
literature is all that is required from those who address
themselves to these pursuits.   He, therefore, who be-
guiles a vacant hour of its tediousness, by some effort of
intellect which captivates the imagination without poi-
soning the morals,—and he who by the exercise of his
art produces forms of beauty which awaken in the mind
that principle of taste which, more than any other
faculty, requires cultivation,—have each bestowed benefits upon the world which may be accurately enough
measured even by the severe limitations of political eco-

nomy; they are profitable labourers and benefactors of their species.

We have entered into these details, principally to show that there are other and higher producers in society than the mere manual labourers. It was an ignorant fashion amongst the mental labourers of other days to despise the physical labourers. They have learnt to know their value; and there should be a reciprocal knowledge. Both classes are working-classes. No one can say that the mental labourers are not workers. They are, we may truly affirm, taken as a class, the hardest workers in the community. No one ever reached eminence in these pursuits without unwearied industry: the most eminent have been universally despisers of ease and sloth, and have felt their highest pleasures in the absorbing devotion of their entire minds to the duties of their high calling. They have wooed Knowledge as a mistress that could not be won without years of unwearied assiduity. The most eminent, too, have been practical men, despising no inquiry, however trifling it might appear to common eyes, and shrinking from no occupation, however tedious, as long as it was connected with their higher duties. The *positive* influence even of the labours of the poets and the artists upon the advance of other labour might be easily shown. In their productions, especially, supply goes before demand, and creates demand. It has been calculated by an American writer, that the number of workmen who have been set in action—paper-makers, printers, binders—by the writings of Sir Walter Scott alone, in all countries, would, if gathered together, form a community that would fill a large town. The Potteries of Etruria, in Staffordshire, could not have existed unless Mr. Wedgwood had introduced into our manufacture of china the forms of Grecian art, bequeathed to us by the taste of two thousand years ago, and thus created a demand which has furnished profitable labour to thousands. But this, as we have already shown, is not the principal way of viewing the influence of science, and literature, and art, upon all other industry. To reduce every labour in art, or literature, or science, to the same standard of value by which manual labour is measured, would be as absurd as the tasteless ignorance of the Spaniards, who applied a rare and valuable antique bust to serve as the weight to a church clock. Any attempt to put the mental labourers upon the same footing of value as the labourers without skill, would be as impossible as it would be mischievous if it were possible ; for in that case production would decline, and ultimately cease altogether, for the fountains of labour with skill would be dried up. Capital must go forward working with accumulation of knowledge ; and fortunately, if the working men adapt themselves to this natural energy of capital, they will themselves become the accumulators of knowledge. Manual labour is only in the highest degree required in the early settlement of a country. When a dense population succeeds to a scattered one, labour with skill is called into action. The counter-control to the absorbing power of capital is the equally absorbing power of skill — for that also is capital. Knowledge is power, because knowledge is property.

CHAPTER XXVII.

In looking back upon the great power of capital working with accumulation of knowledge, we cannot attempt to shut our eyes to the fact that manual labour often suffers for a time by the introduction of a more efficient instrument of production. There are great temporary evils, for instance, on the introduction of a new machine. But in the consideration of such evils we must never forget the principle which we have sought to impress throughout this book,—that the object for which capital works with skill, and the object which machinery effects, is cheapness of production. Machines either save material, or diminish labour, or both. " Which is the cheapest," said the Committee to Joseph Foster, " a piece of goods made by a power-loom, or a piece of goods made by a hand-loom?" He answered, " A power-loom is the cheapest." What, then, is the effect of this reduced cost of production, ultimately, upon the employment of labour? That the manufacture is increased,—that more cloth is consumed,—that the consumer has more money to lay out in cloth, or more money to lay out in other things. We have shown, most distinctly, the effect of the spinning machinery in increasing twenty-fold the number of people engaged in that branch of manufacture. But let us put the circumstance in the words of a person old enough to recollect the precise facts connected with the first introduction of that machinery. The Committee that we have so often mentioned examined Mr. Fielden, who was then a resident at Blackburn, upon this particular subject.

" *Q.* Do you remember what occurred in Lancashire when spinning-factories were first established ?

A. I recollect that period very well.

Q. Were not a very considerable number of persons thrown out of work, and was not there great distress, in consequence of the introduction of machinery, when spinning was done by machinery and not by hand-labour ?

A. Yes, there was a great deal of distress, and much rioting took place at the time.

Q. Persons who had formerly obtained a good living by spinning by hand-labour, were unable to obtain the same wages, and in the same manner, in consequence of the introduction of spinning machinery ?

A. If the description of spinning that was carried on in the neighbourhood of Blackburn is alluded to, that which was done by the hand, the raw cotton was taken out by the weaver, and spun in his own house, and the change was productive of considerable inconvenience in the first instance ; great alarm was created, and some spinning-mills were destroyed at the time ; many persons were thrown out of employment ; but at that time the manufacture of the kingdom was in a very limited state, compared with what it is at present.

Q. Was not the result of the introduction of that machinery an immense increase of the manufacture ?

A. Very great.

Q. And more advantageous wages for a considerably increased number ?

A. Yes, materially so."

It is, we think, self-evident, that if the temporary distress of the hand-spinners, which produced the rioting, and the destruction of spinning-mills here described, had gone on to prevent altogether the manufacture of cotton thread by the spinning machinery, the consumption of cotton cloth would have been little increased, and the number of persons engaged in the manufacture would have been twenty, thirty, or even forty times less than the present number. But there would have been another result. Would the great body of the people of Europe have chosen to wear for many years *dear* cloth instead of *cheap* cloth, that a few thousand spinners might have been kept at their ancient wheels in Lancashire ? Capital can easily shift its place, and invention follows where capital goes before. The people of France, and Germany, and America, would have employed the cheap machine instead of the dear one ; and the people of England would have had cheap cloth instead of dear cloth from thence. We cannot build a wall of brass round our islands ; and the thin walls of prohibitive duties are very easily broken through. A profit of from twenty to thirty per cent. will pour in any given quantity of smuggled goods that a nation living under prohibitive laws can demand. Buonaparte, in the height of his power, passed the celebrated Berlin decree for the exclusion of all English produce from the continent of Europe. But our merchants laughed at him. The whole coast of France, and Holland, and Italy, became one immense receiving place for smuggled goods. If he had lined the whole coast with all the six hundred thousand soldiers that he marched to Russia, instead of a few custom-house officers, he could not have stopped the introduction of English produce. It was against the nature of things that the people who had been accustomed to cheap goods should buy dear ones ; or that they should go without any article, whether of necessity or luxury, whose use had become general. Mark, therefore, if the cotton-spinners of Lancashire had triumphed sixty years ago over Arkwright's machinery, there would not have been a single man, woman, or child of those spinners employed *at all*, within twenty years after that most fatal triumph. The manufacture of cotton would have gone to other countries ; cotton spinning in England would have been at an end. The same thing would have happened if the power-loom, thirty years ago, had been put down by combination ; that is, if the hand-loom weavers had not been as well-informed and as reasonable as we see they are. Mr. Fielden, whose evidence we have already quoted, says, " The introduction of the power-loom, I conceive, will be the cause of saving the manufactures to this kingdom ; without the power-looms the manufactories must be annihilated entirely, for the Americans are making use of the power-loom."

Those who have taken a superficial view of the question of machinery say, that, only whenever there is a greater demand than the existing means can supply, every new discovery in mechanics is a benefit to society, because it gives the means of satisfying the existing wants ; but that, on the contrary, whenever the things produced are sufficient for the consumers, the discovery is a calamity, because it does not add to the enjoyments of the consumers ; it only gives them a better market, which better market is bought at the price of the existence of the producers.

All such reasoning is false in principle, and unsupported by experience. There is no such thing, nor, if machines went on improving for five hundred years at the rate they have done for the last century, could there be any such thing, as a limit to the wants of the consumers. The great mass of facts which we have brought together in this book must have shown, that the cheaper an article of necessity becomes, the more of it is used ; that when the most pressing wants are supplied, and supplied amply by cheapness, the consumer has money to lay out upon new wants ; that when these new wants are supplied cheaply, he goes on again and again to other new wants ; that there are no limits, in fact, to his wants as long as he has any capital to satisfy them. Bear in mind this ; that the first great object of every invention and every improvement is to confer a benefit upon the consumers,—to make the commodity cheap and plentiful. The working man stands in a double character ; he is both a producer and a consumer. But we will be bold to say that the question of cheapness of production is a much more important question to be decided in his favour as a consumer, than the question of dearness of production to be decided in his favour as a producer. The truth is, every man tries to get as much as he can for his own labour, and to pay as little as he can for the labour of others. If a mechanic, succeeding in stopping the machine used in his own trade, by any strange deviation from the natural course of things were to get higher wages for a time, he himself would be the most injured by the extension of the principle. When he found his loaf cost him two shillings instead of one ; when he was obliged to go to the river with his bucket for his supply of water; when his coals cost a guinea a bushel instead of eighteenpence ; when he was told by the hosier that his worsted stockings were advanced from a shilling a pair to five shillings ; when, in fact, the price of every article that he uses should be doubled, trebled, and, in nine cases out of ten, put beyond the possibility of attainment ;—what, we ask, would be the use to him of his advance in wages ?   Let us never forget that it is not for the employment of labourers, but for the benefit of consumers, that labour is employed at all.   The steam-engines are not working in the coal-pits of Northumberland, and the ships sailing from the Tyne to the Thames, to give employment to colliers and to sailors, but to make coals cheap in London.   If the people of London could have the coals without the steam-engines and the ships, it would be better for them, and better for the rest of the world.   If they could get coals for nothing, they would have more produce to exchange for money to spend upon other things ; and the comforts, therefore, of every one of us would be increased.

This increase of comfort, some may say, is a question that more affects the rich than it affects the great mass. This again is a mistake.   The whole tendency of the improvements of the last four hundred years has not only been to lift the meanest, in regard to a great many comforts, far above the condition of the rich four hundred years ago, but absolutely to place them, in many things, upon a level with the rich of their own day.   They are surrounded, as we have constantly shown throughout this book, with an infinite number of comforts and conveniences which had no existence two or three centuries ago ; and those comforts and conveniences are not used only by a few, but are within the reach of almost all men.   Every day is adding something to our comforts.

Our houses are better built—our clothes are cheaper—we have a number of domestic utensils, whose use even was unknown to our ancestors—we can travel cheaply from place to place, and not only travel at less expense, but travel ten times quicker than the richest man could travel two hundred years ago. The bulk of society is not only advancing steadily to the same level in point of many comforts with the rich, but is gaining that knowledge which was formerly their exclusive possession. Let all of us who are producers keep fast hold of that last and best power.

We have endeavoured to show throughout this book that the one great result of machinery, and of every improvement in art, is to lessen the cost of production ; to increase the benefit to the consumer. But it is a most fortunate arrangement of the social state, as we have also shown, that cheap production gives increased employment. The same class of false reasoners who consider that the wants of society are limited, cry out, it is better to have a population of men than of steam-engines. That might be true, if the steam-engines *did* put out the men ; but inasmuch as they increase the productions by which men are maintained, they increase the men. What has increased the population of England nearly tenfold during the last five hundred years, but the improvement of the arts of life, which has enabled more men to live within the land ? There is no truth so clear, that as the productions of industry multiply, the means of acquiring those productions multiply also. The productions which are created by one producer furnish the means of purchasing the productions created by another producer ; and, in consequence of this double production, the necessities of both the one and the other are better supplied. The multiplication of produce multiplies the consumers of produce. There are, probably, upon the average, no more hats made in the year than there are heads to wear them ; but as there are sixteen millions of heads of the British subjects of Queen Victoria, and there were only five millions of the British subjects of Queen Anne, it is self-evident that the hat-makers have three times as much work as they had a century and a quarter ago. What has given the hat-makers three times as much work? The trebling of the population. And what has trebled the population? The trebling of produce—the trebling of the means of maintaining that population.

If the reader has rightly considered the various facts which we have thus presented, he will long before this have come to the conclusion, that it is for the general interests of society that every invention, which has a tendency to diminish the cost of production, shall have the most perfect freedom to go forward. He will also have perceived, that the exercise of this natural right, this proud distinction, of man, to carry on the work of improvement to the fullest extent of his capacity and knowledge, can never be wholly stopped, however it may be opposed. It may be suspended by the ignorance of a government—it may be clamoured down by the prejudice of a people; but the living principle which is in it can never be destroyed. To deny that this blessing, as well as many other blessings which we enjoy, is not productive of any particular evil, would be uncandid and unwise. Every change produced by the substitution of a perfect machine instead of an imperfect one, of a cheap machine instead of a dear one, is an inconvenience to those who have been associated with the imperfect and the dear machines. It is a change that more or less affects the interests of capitalists as well as of workmen. In a commercial country, in a highly civilized community, improvement is hourly producing some change which affects some interests. Every new pattern which is introduced in hardware deranges for a moment the interests of the proprietors of the old moulds. Every new book, upon any specific subject upon which books have formerly been written, lessens the value of the copyright of those existing books.—What then? Is every improvement, which thus produces a slight partial injury, to be discountenanced, because of this inevitable condition which we find at every step in the march of society? Or rather, ought we not to feel that every improvement brings healing upon its wings, even to those

for whom it is a momentary evil;—that if it displaces
their labour or their capital for a season, it gives new
springs to the general industry, and calls forth all labour
and all capital to higher and more successful exertions?

At every advance which improvement makes, the partial and temporary evils of improvement are more and
more lessened. In the early stages of social refinement,
when a machine for greatly diminishing labour is for the
first time introduced, its effects in displacing labour for
an instant may be seen in the condition of great masses
of people. It is the first step which is the most trying.
Thus, when printing superseded the copiers of books
by writing, a large body of people were put out of employ;—they had to seek new employ. It was the same
with the introduction of the spinning machinery,—the
same with the power-loom. It would be presumptuous
to say that no such great changes could again happen in
any of the principal branches of human industry; but it
may be said, that the difficulty of introducing more expeditious and cheaper modes of manufacture is daily
increasing. The more machines are multiplied, that is,
the more society approaches towards perfection, the less
room is there for those great inventions which change
the face of the world. We shall still go on improving,
doubtless; but ingenuity will have a much narrower
range to work in. It may perfect the machines which
we have got, but it will invent fewer new machines.
And who can doubt, that the nearer we approach to this
state, the better will it be for the general condition of
mankind? Who can doubt whether, instead of a state
of society where the labourers were few and wretched,
wasting human strength, unaided by art, in labours which
could be better performed by wind, and water, and
steam,—by the screw and the lever,—it would not be
better to approach as nearly as we can to a state of society where the labourers would be many and lightly
tasked, exerting human power in its noblest occupation,
that of giving a direction by its intelligence to the mere
physical power which it had conquered? Surely, a nation so advanced as to apply the labour of its people to

occupations where a certain degree of intelligence was required, leaving all that was purely mechanical to machines and to inferior animals, would produce for itself the greatest number of articles of necessity and convenience, of luxury and taste, at the cheapest cost. But it would do more. It would have its population increasing with the increase of those productions; and that population employed in those labours alone which could not be carried on without that great power of man by which he subdues all other power to his use,—his reason.

CHAPTER XXVIII.

THE reader will remember that when the fur-traders refused to advance to John Tanner a supply of blankets for his winter consumption, he applied himself to make garments out of moose-skins. The skin was ready manufactured to his hands when he had killed and stripped the moose ; but still the blanket brought from England across the Atlantic was to him a cheaper and a better article of clothing than the moose-skin which he had at hand ; and he felt it a privation when the trader refused it to him upon the accustomed credit. It never occurred to him to think of manufacturing a blanket ; although he was in some respects a manufacturer. He was a manufacturer of sugar, amongst the various trades which he followed. He used to travel about the country till he had found a grove of maple-trees ; and here he would sit down for a month or two till he had extracted sugar from the maples. Why did he not attempt to make blankets ? He had not that Accumulated Knowledge, and he did not work with that DIVISION OF LABOUR, which are essential to the manufacture of blankets—both of which principles are carried to their highest perfection when capital enables the manufacture of woollen cloth, or any other article, to be carried forward upon a large scale.

Let us endeavour to trace what accumulations of skill, and especially what divisions of employment, were necessary to enable Tanner to clothe himself with a piece of woollen cloth. We shall not stop to inquire whether the skill has produced the division of employment, or the division of employment has produced the skill. It is sufficient for us to show, that the two principles are in joint operation, unitedly carrying forward the business of production in the most profitable manner. It is enough for us to know, that where there is no skill there is no division of employment, and where there is no division of employment there is no skill. Skill and division of employment are inseparably wedded. If they could be separated, they would in their separation cease to work profitably. They are kept together by the constant energy of capital, devising the most profitable direction for labour.

Before a blanket can be made, we must have the material for making a blanket. Tanner had not the material, because he was not a cultivator. Before wool can be grown there must be, as we have shown, appropriation of land. When this appropriation takes place, the owner of the land either cultivates it himself, which is the earliest stage in the division of agricultural employment,—or he obtains a portion of the produce in the shape of corn or cattle, or in a money payment. Hence a tenantry. But the tenant, to manufacture wool at the greatest advantage, must possess capital, and carry forward the principle of the division of employment by hiring labourers. We use the word *manufacture* of wool advisedly; for all farming processes are manufacturing processes, and invariably reduce themselves to change of form, as all commercial processes reduce themselves to change of place. If the capital of the farmer is sufficient to enable him to farm upon a large scale, he divides his labourers; and one becomes a shepherd, one a ploughman,—one sows the ground, and one washes and shears the sheep, more skilfully than another. If he has a considerable farm, he divides his land, also, upon the same principle, and has pasture, and arable, and rotation of crops. By these divisions he is enabled to manufacture wool cheaper than the farmer upon a small scale, who employs one man to do everything, and has not a proper proportion of pasture and arable, or a due rotation of crops. At every division of employment skill must be called forth in a higher perfection than when two or more employments were joined together; and that the chief director of the skill, the capitalist himself, or farmer, must require more skill to make all the parts which compose his manufactory work together harmoniously.

But we have new divisions of employment to trace before the wool can be got to the manufacturer. These employments are created by what may be called the *local* division of labour. It is convenient to rear the sheep upon the mountains of Wales, because there the short and thymy pastures are fitted for the growth of wool. It is convenient to manufacture the wool into cloth at Leeds, because coals are there at hand to give power to the steam-engines, with which the manufacture is carried on. The farmer in Wales, and the manufacturer of cloth at Leeds, must be brought into connexion. In the infancy of commerce one or both of them would make a journey to establish this connexion; but the cost of that journey would add to the cost of the wool, and therefore lessen the consumption of woollen cloth. The division of employment goes on to the creation of a wool-factor, or dealer in wool, who either purchases directly from the grower, or sells to the manufacturer for a commission from the grower. The grower, therefore, sends the wool direct to the factor, whose business it is to find out what manufacturer is in want of wool. If the factor did not exist, the manufacturer would have to find out, by a great deal of personal exertion, what farmer had wool to sell; or the farmer would have to find out, with the same exertion, what manufacturer wanted to buy wool. The factor receives a commission, which the seller and buyer ultimately unite in paying. They co-operate to establish a wool-factor, just as we all co-operate to establish a postman; and just as the postman who delivers a number of letters to a great many individuals, does that service at little more cost to all, than each individual would pay for the delivery of a single letter, so does the wool-factor exchange the wool between the grower and the manufacturer, at little more cost to a large number of the growers who employ him, than each would be obliged to pay in expenses and loss of time to travel from Wales to Leeds to sell his wool.

We have, however, a great many more divisions of employment to follow out before the wool is conveyed from Wales to Leeds. If the packs are taken on shipboard, and carried down the Mersey to Liverpool, we have all the variety of occupations, involving different degrees of skill, which make up the life of a mariner; if they go forward upon the railroad to Manchester, we have all the higher degrees of skill involved in their transport which belong to the business of an engineer: and if they finally reach Leeds by canal, we have another division of labour that adjusts itself to the management of boats in canals. But the ship, the railroad, the canal, which are created by the necessity of transporting commodities from place to place, have been formed after the most laborious exercise of the highest science, working with the greatest mechanical skill; and they exist only through the energy of prodigious accumulations of capital, the growth of centuries of patient and painful labour and economy.

We have at length the wool in a manufactory at Leeds. The very employment of the circulating capital which has brought it there has produced great division of labour, engaged in carrying forward and maturing the complicated operations of credit. We shall not dwell upon the machinery by which the wool is converted into cloth with the greatest saving of time and material, and with a perfection which no labour of the hand, unassisted by science, could ever attain. We shall call attention only to the great division of employments into which the process of manufacturing wool into cloth has branched, under the guidance of large capital, striving to accomplish production at the cheapest rate.

The first class of persons who prepare the wool, are the sorters and pickers. It is their business to separate the fine from the coarse locks, so that each may be suited to different fabrics. There is judgment required, which could not exist without division of labour; and the business, too, must be done rapidly, or the cost of sorting

and picking would outweigh the advantage.   The second
principal operation is scouring.   Here the men are constantly employed in washing the wool, to free it from all
impurities.   It is evident that the same man could not
profitably pass from the business of sorting to that of
scouring, and back again,—from dry work to wet, and
from wet to dry.   When the wool is out of the hands of
the scourers it comes into those of the dyers, who colour
it with the various chemical agents applied to the manufacture.   The carders next receive it, who tear it with
machines till it attains the requisite fineness.   From the
carders it passes to the slubbers, who form it into tough
loose threads ; and thence to spinners, who make the
threads finer and stronger.   There are sub-divisions of
employment which are not essential for us to notice, to
give an idea of the great division of employment, and
the consequent accumulation of peculiar skill, required
to prepare wool to be made into yarn, to be made into
woollen cloth.

The next stages in the manufacture are the spinning,
the warping, the sizing, and the weaving.   These are
all distinct operations, and are all carried forward with
the most elaborate machinery, adapted to the division of
labour which it enforces, and by which it is enforced.

But there is a great deal still to be done before the
cloth is fit to be worn.   The cloth, now woven, has to
be scoured as the wool was.   There is a subsequent
process called burling, at which females are constantly
employed.   The boiling and milling come next, in which
the cloth is again exposed to the action of water, and
beaten so as to give it toughness and consistency.
Dressers, called giggers, next take it in hand, who also
work with machinery upon the wet cloth.   It has then
to be dried in houses where the temperature is sometimes as high as 130 degrees, and where the men work
almost naked.   It is evident that the boilers and dressers
could not profitably work in the dry-houses ; and that
there must be division of employment to prevent those
sudden transitions which would destroy the human frame
much more quickly than a regular exposure to cold or

heat, to damp or dryness. The cloth must be next cropped or cut upon the face, to remove the shreds of wool which deform the surface in every direction. When cut, it has to be brushed dry by machinery, to get out the croppings which remain in its texture. This done, it is dyed in the shape of cloth, as it was formerly dyed in the shape of wool. Then come a variety of processes, to increase the delicacy of the fabric :—singeing, by passing the cloth within a burning distance of red-hot cylinders ; frizing, to raise a nap upon the cloth ; glossing, by carrying over it heavy heated plates of iron ; pressing, in which operation of the press red-hot plates are also employed ; and drawing, in which men, with fine needles, draw up minute holes in the cloth when it has passed through the last operation. Then comes the packing ; and after all these processes it must be bought by a wholesale dealer, and again by a retailer, before it reaches the consumer. Between the growth of the fleece of wool, and the completion of a coat by a skilful tailor,— who, it is affirmed, puts five-and-twenty thousand stitches into it,—what an infinite division of employments! what inventions of science! what exercises of ingenuity ! what unwearied application! what painful, and too often unhealthy labour ! And yet if men are to be clothed well and cheaply, all these manifold processes are not in vain ; and the individual injury in some branches of the employ is not to be compared to the suffering that would ensue if cloth were not made at all, or if it were made at such a cost that the most wealthy only could afford to wear it. But for the accumulation of knowledge, and the division of employments, engaged in the manufacture of cloth, and set in operation by large capital, we should each be obliged to be contented with a blanket such as John Tanner desired ; and very few indeed would even obtain that blanket: for if skill and division of labour were not to go on in one branch, they would not go on in another, and then we should have nothing to give in exchange for the blanket. The individual injury to health, also, produced by the division of labour, is not so great, upon the average, as if there were no division. All the returns of human life in this country show an extremely little difference in the effect upon life, even of what we consider the most unhealthy trades ; and this proceeds from that extraordinary power of the human body to adapt itself to a habit, however apparently injurious, which is one of the most beautiful evidences of the compensating principle which prevails throughout the moral world.

CHAPTER XXIX.

HAVING traced the operation of the division of labour,—
and one instance, such as that of the manufacture of
cloth, will do for an example of the wonders it effects,—
we may not improperly devote a few words to a view of
the influence of the great extension of the principle upon
the condition of the operatives.

In France, which, as a commercial and manufacturing
country, is considerably behind the advance of England,
it is a common practice, in many villages and small towns,
for the weavers to make the looms and other implements
of their trade. In the fifteenth century, in the same
country, before an apprentice could be admitted to the
privilege of a master-weaver, it was not only necessary
for him to prove that he understood his trade as a weaver,
but that he was able to construct all the machines and
tools with which he carried on his craft. Those who
know anything of the business of weaving will very readily come to the conclusion that the apprentice of the
fifteenth century, whose skill was put to such a proof,
was both an indifferent weaver and an indifferent mecha-
nician ;—that in the attempt to unite two such opposite
trades, he must have excelled in neither ;—and that in
fact the regulation was one of those monstrous violations
of the freedom of industry, which our ancestors chose to
devise for the support of industry.

Carrying the principle of a division of labour to the
other extreme point, we have seen that a vast number of
persons are engaged in the manufacture of a piece of
cloth, who, if individually set to carry the workmanship
of that piece of cloth through all its stages, would be

L 2

utterly incompetent to produce it at all, much less to
produce it as durable and beautiful as the cloth which
we all daily consume. How would the sorter of the
wool, for example, know how to perform the business
of the scourer, or of the dyer, or of the carder ?— or the
carder that of the spinner or the weaver ?—or the weaver
that of the miller, or boiler, or dyer, or brusher, or cut-
ter, or presser ? We must be quite sure that, if any
arbitrary power or regulation, such as compelled the
weaver of the fifteenth century to make his own loom,
were, on the other hand, to compel a man engaged in any
one branch of the manufacture of woollen cloth to carry
that manufacture through all its stages, the production of
cloth would be utterly suspended ; and that the workmen
being incompetent to go on, the wages of the workmen
could no longer be paid :—for the wages of labour are
paid by the consumer of the products of labour, and here
there would be nothing to consume.

The great principle, therefore, which keeps the division of labour in full activity is, that the principle is
necessary to production upon a scale that will maintain
the number of labourers engaged in working in the
cheapest, because most economical manner, through the
application of that mode of working. The labourers,
even if the principle were injurious to their individual
prosperity and happiness, which we think it is not, could
not dispense with the principle, because it is essential to
economical production ; and if dear production were to
take the place of economical production, there would be
a proportionately diminished demand for products, and
a proportionate diminution of the number of producers.

The same laws of necessity which render it impossible
for the working men to contend against the operation of
the division of labour,—even if it were desirable that
they should contend against it, as far as their individual
interests are concerned,—render it equally impossible
that they should contend against the operation of accumulation of knowledge in the direction of their labour.
The mode in which accumulation of knowledge influences
the direction of their labour is, that it furnishes mecha-

nical and chemical aids to the capitalist for carrying on the business of production. The abandonment of those mechanical and chemical aids would suspend production, and not in the slightest degree increase, but greatly diminish, and ultimately destroy, the power of manual labour, seeking to work without those mechanical and chemical aids. The abandonment of the division of labour would work the same effects. There would be incomparably less produced on all sides ; and the workmen on all sides, experiencing in their fullest extent the evils which result from diminished production, would all fall back in their condition, and day by day have less command of the necessaries and comforts of life, till they sank into utter destitution.

We dwell principally on the effects of accumulation of knowledge and division of labour on the working man as a consumer, because it is the immediate object of this volume to consider such questions with reference to production. But the condition of the working man as a producer is, taking the average of all ranks of producers, greatly advanced by the direction which capital gives to labour, by calling in accumulation of knowledge and division of labour. If the freedom of labour were not established upon the same imperishable basis as the security of property, we might, indeed, think that it was a pitiable thing for a man to labour through life at one occupation, and believe that it was debasing to the human intellect and morals to make for ever the eye of a needle, or raise a nap upon woollen cloth. The Hindoos, when they instituted their *castes*, which compelled a man to follow, without a possibility of emerging from it, the trade of his fathers, saw the general advantage of the division of labour ; but they destroyed the principle which could make it endurable to the individual. They destroyed the Freedom of Industry. " To limit industry or genius, and narrow the field of individual exertion by any artificial means, is an injury to human nature of the same kind as that brought on by a community of possessions. Where there is no stimulus to industry, things are worst ; where industry is circumscribed, they cannot prosper ; and are then only in a healthy state, when every avenue to personal advantage is open to every talent and disposition.   A state of equality is an instance of the first case ; the division of the people into castes, as among the Ancient Egyptians, and still among the Hindoos, of the second.   This division has been considered by all intelligent travellers as one powerful cause of the stationary character of the inhabitants of that country : and the effect would have been still more pernicious, if time or necessity had not introduced some relaxation into the rigorous restrictions originally established, and so ancient as to be attributed to Siva.   As long, however, as the rule is generally adhered to, that a man of a lower class is restricted from the business of a higher class, so long, we may safely predict, India will continue what it is in point of civilization.   An approach to the same effect may be witnessed in the limitation of honours, privileges, and immunities in some countries of Europe."*

When a man can pass from occupation to occupation, as he is now allowed by the laws of this country to do, and as he will be completely allowed when the rights of industry are better understood by the votaries of industry, the division of labour will not press severely upon any man.   In those manufactures and trades where the division of labour is carried to the greatest extent, such as the cotton and silk trades, workmen readily change from one branch to the other, without molestation, and without any great difficulty of adapting themselves to a new occupation.   The simpler the process in which a workman has been engaged—and every process is rendered more simple by the division of labour—the easier the transition : and the principal quality which is required to make the transition is, that stock of general knowledge which the division of labour enables a man to attain ; and which, in point of fact, is attained in much higher perfection in a large manufactory, than in that rude state where one man is more or less compelled to do everything for his

* Sumner's Records of the Creation.

body, and therefore has no leisure to do anything for his mind. There are evils, undoubtedly, in carrying the division of labour to an extreme point, but we think that those very evils correct themselves, because they destroy the great object of the principle, and give imperfect instead of perfect production. The moral evils which some have dreaded may assuredly be corrected by general education, and in fact are corrected by the union of numbers in one employment. What sharpens the intellect ought, undoubtedly, to elevate the morals; and, indeed, it is only false knowledge which debases the morals. Knowledge and virtue we believe are the closest allies; and wisdom is the fruit of knowledge and virtue.

The same principles as to the course which the division of labour should lead the labourer to pursue, apply to the higher occupations of industry. No man of learning has ever very greatly added to the stock of human knowledge, without devoting himself, if not exclusively, with something like an especial dedication of his time and talents, to one branch of science or literature. In the study of nature we have the mathematician, the astronomer, the chemist, the botanist, the zoologist, and the physician engaged, each in his different department. In the exposition of moral and political truths, we have the metaphysician, the theologian, the statesman, the lawyer, occupied each in his peculiar study or profession. A mental labourer, to excel in any one of these branches, must know something of every other branch. He must direct indeed the power of his mind to one department of human knowledge; but he cannot conquer that department without a general, and, in many respects, accurate knowledge of every other department. The same principle produces the same effects, whether applied to the solution of the highest problem in geometry, or the polishing of a pin. The division of labour must be regulated by the acquisition of general knowledge.

But the division of labour in carrying forward the work of production is invariably commanded, because it is perfected, by the union of forces or co-operation. The process of manufacturing a piece of woollen cloth, which we have described, is carried on by division of labour, and by union of forces, working together. In fact, if there were not that ultimate co-operation, the division of labour would be not only less productive than labour without division, but it would not be productive at all. The power of large capital is the power which, as society is arranged, compels this division of parts for the more complete production of a whole. A large cloth manufactory, as we have seen, exhibits itself to the eye chiefly in the division of labour; but all that division ends in a co-operation for the production of a piece of cloth. A ship, with five hundred men on board, each engaged in various duties and holding different ranks, is an example of the division of labour; but the division ends in a co-operation to carry the ship from one port to another, and, if it be a ship of war, to defend it from the attacks of an enemy. Those who would direct the principle of co-operation into a different channel, by remodelling society into large partnerships, do not, because they cannot, depart, in the least degree, from the principles we have laid down. They must have production, and therefore they must have division of labour : the division of labour involves degrees of skill; the whole requires to be carried on with accumulation of former labour or capital, or it could not exist. The only difference proposed is, that the labourers shall be the capitalists, and that each shall derive a share in the production, partly from what now is represented as his profits as a capitalist, and partly from what is represented as his wages as a labourer; but that all separate property shall be swallowed up in joint property. But we mention this subject here to show that even those who aspire to remodel society cannot change the elements with which it is now constructed, and must work with the same principles, however different may be the names of those principles, and however varied in their application. This is in favour even of the ultimate success of the principles of co-operation, if they should be found practically to work for the increase of the happiness of mankind, which would not be effected by equalizing the distribution of wealth, if, at the same time, its production were materially checked. This view of the subject goes to show that no sudden or violent change is necessary. In many things society has always acted on the principles of co-operation. As civilization extends, the number of instances has hitherto increased ; and if there is no natural maximum to the adoption of these principles (which remains to be seen), men may gradually slide more and more into them, and realize all sane expectations, without any reconstruction of their social system,—any pulling down and building up again of their morals or their houses.

It is this union of forces which, whether it prevail in a single manufactory, in a manufacturing town viewed in connexion with that manufactory, in an agricultural district viewed in connexion with a manufacturing town, in a capital viewed in connexion with both, in a kingdom viewed in connexion with all its parts, and in the whole world viewed in connexion with particular kingdoms ;—it is this union of forces which connects the humblest with the highest in the production of utility. The poor lad who tends sheep upon the downs, and the capitalist who spends hundreds of thousands for carrying forward a process to make the wool of these sheep into cloth, though at different extremities of the scale, are each united for the production of utility. The differences of power and enjoyment (and the differences of enjoyment are much less than appear upon the surface) between the shepherd boy and the great cloth-manufacturer, are apparently necessary for the end of enabling both the shepherd boy and the capitalist to be fed, and clothed, and lodged, by exchanges with other producers. They are also necessary for keeping alive that universal, and, therefore, as it would appear, natural desire for the improvement of our condition, which, independently of the necessity for the satisfaction of immediate wants, more or less influences the industry of every civilized being as to the hopes of the future. It is this union which constitutes the real dignity of all useful employments, and may make the poorest labourer feel that he is advancing the welfare of mankind as well as the richest capitalist ; and that, standing upon the solid foundation of free exchange, the rights of the one are as paramount as the rights of the other, and that the rights of each have no control but the duties of each. We believe that the interests of each are also inseparably united, and that the causes which advance or retard the prosperity of each are one and the same. We have shown that capital and labour must work, to work profitably, with interchangeable freedom and security; that when the security or the freedom of either is impeded, production languishes; that in their union of mutual freedom and security capital must of necessity work upon the principles which best advance production, and that therefore capital must work with machinery with division of labour and accumulation of knowledge; and, finally, that when production is carried forward by these powerful agencies, the situation of every consumer is bettered, by the increased command of the necessaries and comforts of life which cheap production places within his reach. It remains for us to show the particular mode in which capital, working with freedom and security, stimulates production; and therefore calls forth labour by assuming a circulating form, which represents the consumable commodities destined for the maintenance of labour.

CHAPTER XXX.

THERE is one other element in the production of wealth which we have incidentally noticed in various parts of this volume, but which is such an essential ingredient in the carrying forward exchanges, without which no wealth can exist, that we must state its operations formally, to arrive at a full knowledge of all the circumstances which determine production.

When the prodigal, whose course we have described in our fifth chapter, obtained an exchange of consumable commodities for his labour, he received those commodities in the shape of " a few pence and a small gratuity of meat and drink." In this transaction we find the circumstances which represent every exchange of labour for capital. The prodigal wanted meat and drink, and he gave labour in exchange for meat and drink ; the capitalist wanted the produce of labour—he wanted a new value bestowed upon his coals by labour—and he gave meat and drink in exchange for the labour which the prodigal had to give. But the prodigal wanted something beyond the meat and drink which was necessary for the supply of the day. He had other immediate necessities beyond food ; and he had determined to accumulate capital. He therefore required " a few pence " in addition to the " meat and drink." The capitalist held that the labour performed had conferred a value upon his property, which would be fairly exchanged for the pence in addition to the food ; and he gave, therefore, in exchange, that portion of his capital which was represented by the money and by the food. This blending of one

sort of consumable commodity, and of the money which
represented any other consumable commodity which the
money could be exchanged with, was an accident arising
out of the peculiar circumstances in which the prodigal
happened to be placed. In ordinary cases he would have
received the money alone ; that is, he would have received a larger sum of money to enable him to exchange
for meat and drink, instead of receiving them in direct
payment. It is clear, therefore, that as the money represented one portion of the consumable commodities
which were ready to pay for the labour employed in
giving a new value to the coals, it might represent
another portion—the meat, for instance, without the
drink ; or it might represent all the consumable commodities, meat, drink, lodging, clothes, fuel, which that
particular labourer might want ; and even represent the
accumulation which he might extract out of his self-
denial as to the amount of meat, drink, lodging, clothes,
and fuel which he might require as a consumer.

We have already mentioned that if the labourer, whose
story we have told, had received a portion of the coals
upon which he had conferred a new value in exchange
for the labour which produced that value, he would have
been paid in a way that would have naturally arisen out
of a rude state of society. But he would have been paid
in a way very unfavourable for production. It would
have required a new labour before the coals could have
procured him the meat, and drink, and lodging of which
he had an instant want ; and he therefore must have received a larger portion of coals to compensate for his new
labour, or otherwise his labour must have been worse
paid. There would have been unprofitable labour, whose
loss must have fallen somewhere,—either upon the capitalist or the labourer in the first instance, but upon both
ultimately, because there would have been less production.
All the unprofitable labour employed in bringing the exchange of the first labour for capital to maturity, would
have been so much power withdrawn from the efficiency
of the next labour to be performed ; and therefore production would have been impeded to the extent of that

unprofitable labour. The same thing would have happened if, advancing a step forward in the science of exchange, the labourer had received an entire payment in meat and drink, instead of part meat and drink, and part money. Wanting lodging, he would have had to seek a person who wanted meat and drink in exchange for lodging, before he could have obtained lodging. But he had a few pence,—he had money. He had a commodity to exchange that he might divide and sub-divide as long as he pleased, whilst he was carrying on an exchange; that is, he might obtain as much lodging as he required for an equivalent portion of his money. If he kept his money, it would not injure by keeping as the food would. He might carry it from place to place more easily than he could carry the food. He would have a commodity to exchange, whose value could not be made matter of dispute, as the value of meat and drink would unquestionably have been. This commodity would represent the same value, with little variation, whether he kept it a day, or a week, or a month, or a year; and therefore would be the only commodity whose retention would advance his design of accumulating capital, with certainty and steadiness. It is evident that a commodity possessing all these advantages must have some intrinsic qualities which all exchangers would recognise—that it must be a standard of value—at once a commodity possessing real value, and a measure of all other values. This commodity exists in all commercial or exchanging nations in the shape of coined metal. The metal itself possesses a real value, which represents the labour employed in producing it; and, in the shape of coin, represents also a measure of other value, because the value of the coin has been determined by the sanction of some authority which all admit. That authority is most conveniently expressed by a Government, as the representative of the aggregate power of society. The metal itself, unless in the shape of coined money, would not represent a definite value; because the metal might be depreciated in value by the admixture of baser or inferior metals, unless it bore the impress of authority to determine its value. The exchangers of the metal for other articles of utility could not, without great loss of labour, be constantly employed in reducing it to the test of value, even if they had the knowledge requisite for so ascertaining its value.   In Greece, a piece of gold in the rude times was stamped with the figure of an ox, to indicate that it would exchange for an ox.   In modern Europe, a piece of gold, called a sovereign, represents a certain weight in gold uncoined, and the Government stamp indicates its purity. A sovereign purchases so many pounds weight of an ox, and a whole ox purchases so many sovereigns.   The great use of the coined metal is to save labour in exchanging the ox for other commodities.   The money purchases the ox, and a portion of the ox again purchases some other commodity, such as a loaf of bread from the baker, who obtains a portion of the ox through the medium of the money, which is a standard of value between the bread and the beef.

But there is another and a last step, in carrying forward the business of exchange with ease to all exchangers, and with consequently less labour and increased production.   That step is Credit.   Credit, upon a large scale, arose from the difficulty of transmitting coined money from place to place, and particularly from one country to another ; and hence the invention of bills of exchange. A bill of exchange is an order by one person on another, to pay to a specified person, or his order, a sum of money specified, at a certain time and a certain place.   It is evident that the bill of exchange travels as much more conveniently than a bag of money, as the bag of money travels more conveniently than the goods which it represents.   For instance, a box of hardware from Birmingham might be exchanged for a case of wine from Bourdeaux, by a direct barter between the tradesman at Birmingham and the tradesman at Bourdeaux ; but this sort of operation must be a very limited one.   Through the agency of merchants, the hardware finds its way to Bourdeaux, and the wine to Birmingham, without any direct exchange between either place, or without either having more of the commodity wanted than is required by the market,—that is, the supply proportioned to the demand of each town. Through the division of labour, the merchant who exports the hardware to Bourdeaux, and the merchant who exports the wine from Bourdeaux, are different people ; and there are other people engaged in carrying on other transactions at and with Bourdeaux, with whom these merchants come in contact. When, therefore, the merchant at Bourdeaux has to pay for the hardware in England, he obtains a bill of exchange from some other merchant who has to receive money from England, for the wine which he has sent there. And thus not only is there no direct barter between the grower of the wine and the manufacturer of the hardware, but the wine and the hardware are each paid for without any direct remittance of coined money from France to England, but by a transfer of the debt due from one person to another in each country. By this transfer, the transaction between the buyer and the seller is at once brought to maturity ; and by this operation the buyer and seller are each benefited, because the exchange which each desires is rendered incomparably more easy, because more speedy and complete. The same principle applies to transactions between commercial men in the same country. The order for payment, which stands in the place of coined money in one case, is called a Foreign bill of exchange ; in the other, an Inland bill of exchange.

The operation of credit in a country whose industry is in an advanced state of activity, is extended over all its commercial transactions, by the necessity of obtaining circulating capital for the carrying forward the production of any commodity, from its first to its last stages. A manufacturer has a large sum expended in workshops, warehouses, machinery, tools. This is called his plant, or fixed capital. He has capital invested also in the raw material which he intends to convert into some article of utility. He works up his raw material ; he makes advances for the labour required in working it up. The article is at length ready for the market. The wholesale dealer, who purchases of the manufacturer, sells to a re-

tailer, who is in the habit of buying upon credit, long or
short, because the article remains a certain time in his
hands before it reaches the consumer, who ultimately
pays for it. From the time when a fleece of wool is
taken from the sheep's back in Wales, till it is purchased
in the shape of a coat in London, there are extensive
outlays in every department, which could not be carried
on steadily unless there were facilities of credit from
one person concerned in the production to another person concerned in the production,—the whole credit
being grounded upon the belief that the debt contracted
in so many stages will be repaid by the sale of the cloth
to the consumer. The larger operations of this credit
are represented by bills of exchange, or engagements to
pay at a given date ; and these bills being converted into
cash by a banker, furnish a constant supply of consumable commodities to all parties concerned in advancing the production, till the produce arrives in the hands
of the consumer. To judge of the extent to which
credit is carried in this country, it is only necessary to
mention, that five millions sterling are daily paid in bills
and cheques by the London bankers alone. The credit,
undoubtedly, if conducted upon fair principles, represents some capital actually in existence, and therefore
does not really add to the accumulation or capital of the
producers. But it enables men in trade at once to have
stock and circulating capital—to use their houses and
shops, and manufactories and implements ; and to give,
at the same time, a security to others upon that fixed
capital. This process is, as it were, as if they coined
that fixed capital. The credit, which is rendered, as
secure as possible in all its stages by the accumulating
securities of the drawer, acceptor, and endorsers of a
Bill of Exchange, brings capital into activity,—it carries
it directly to those channels in which it may be profitably
employed,—it conducts it to those channels by a systematic mode of payment for its use, which we call interest,
or discount ;—and it therefore carries forward accumulation to its highest point of productiveness.

If the reader will turn to the passage in our third

chapter, where Tanner describes the refusal of the traders to give him credit, he will see how capital, advanced upon credit, sets industry in motion. The Indians had accumulated no store of skins to exchange for the trader's store of guns, ammunition, traps, and blankets. The trader, although he possessed the articles which the Indians wanted, refused to advance them upon the usual credit, and they were consequently as useless to the Indians as if they had remained in a warehouse at Liverpool or Glasgow. When the credit was taken away from the Indians, they could no longer be exchangers. Their own necessities for clothing were too urgent to enable them to turn their attention from that supply to accumulate capital for exchange, after the winter had passed away. They hunted only for themselves. The trader went without his skins, and the Indians without their blankets. Doubtless, the keenness of commercial activity soon saw that this state of things was injurious even to the more powerful party, for the accustomed credit was presently restored to the Indians. It was the only means by which that balance of power could be quickly restored which would enable the parties again to become exchangers. Every exchange presupposes a certain equality in the exchangers : and credit, therefore, from the capitalist to the non-capitalist, must, in many cases, be the first step towards any transaction of mutual profit. If the Indians had adopted the resolution of Tanner, to do without the blankets for the winter, and had substituted the more imperfect clothing of skins,—and if the traders had persevered in their system of refusing credit, that is, of advancing capital, —the exchange of furs must have been suspended, until, by incessant industry and repeated self-denial, the Indians had become capitalists themselves. They probably, after a long series of laborious accumulations, might have done without the credit—that is, have not consumed the goods which they received before they were in a condition to give their own goods as equivalents ; and then, as it usually happens in the exchanges of civilized society, they would have ensured a higher

reward for their labour. The credit rendered the labour
of the Indians less severe, inasmuch as it allowed them
to work with the aid of the accumulations of others, instead of with their own accumulations. But it doubtless gave the traders advantage, and justly so, in the
terms of the exchange. If the Indians had brought
their furs to the mart where the dealers had brought
their blankets, there would have been exchange of capital for capital. As the Indians had not accumulated
any furs, and were only hoping to accumulate, there
was, on the part of the white traders, an advance of a
present good for a remote equivalent. The traders had
doubtless suffered by the casualties which prevented the
Indians completing their engagements. They made a
sudden, and therefore an unjust change in their system.
The forbearance of the Indians shows their respect for
the rights of property, and their consequent appreciation
of their own interests. They might, possessing the
physical superiority, have seized the blankets and ammunition of the traders. If so, their exchanges would
have been at an end ; the capital would have gone to
stimulate other industry ; the Indians would have
ripped up the goose with the golden eggs.

It is easy to see that the employment of capital,
through the agency of credit, in all the minute channels
of advanced commerce, must wholly depend upon the
faith which one man has in the stability and the honesty of
another ; and also upon the certainty of the protection
of the laws which establish security of property, to enforce the fulfilment of the contract. It is of the highest
importance, therefore, for commercial men to maintain
their credit unimpeached, and to uphold, also, those laws
and institutions which enable credit generally to go forward without obstruction. Commercial men know how
easily credit is destroyed by individual guilt or imprudence ; and how easily it is contracted, generally, by a
combination of circumstances over which an individual,
apart from a nation, has no control. The instant that
any circumstances take place which weaken the general
confidence in the security of property, credit is with-

drawn. The plant remains—the tools and warehouses stand—the shops are open ; but production languishes-- labour is suspended. The stocks of consumable com- modities for the maintenance of labour may still in part exist, but they do not reach the labourer through the usual channels. Then men say, and say truly, con- fidence is shaken ;—the usual relations of society are disturbed. Capital fences itself round with prudence— hesitates to go on accumulating—refuses to put its ex- istence in peril—withdraws in great part from pro- duction—

"Spreads its light wings, and in a moment flies."

CHAPTER XXXI.

In looking back upon all the various circumstances which we have exhibited as necessary for carrying industry to the greatest point of productiveness, we think that we must have established satisfactorily that the two great elements which concur in rendering labour in the highest degree beneficial, are, 1st, the accumulated results of past labour, and, 2nd, the contrivances by which manual labour is assisted,—those contrivances being derived from the accumulations of knowledge. Capital and skill, therefore, are essential to the productive power of labour. The different degrees in which each possesses capital and skill make the difference between an English manufacturer and a North American savage; and the less striking gradations in the productive power of the English manufacturer of the present time, and the English manufacturer of five hundred years ago, may be all resolved into the fact that the one has at his command a very large amount of capital and skill, and that the other could only command a very small amount of the same great elements of production.

We think, also, that we have shown that the accumulation of former labour in the shape of tangible wealth, and the accumulation of former labour in the shape of the no less real wealth of knowledge, are processes which go on together, each supporting, directing, and regulating the other. Knowledge is the offspring of some leisure resulting from a more easy supply of the physical wants; and that leisure cannot exist unless capital exists; which allows some men to live upon former accumulations. Capital, therefore, may be said to be the parent of skill, as capital and skill united are the encouragers and directors of profitable labour.

We have shown, that the only foundation of accumulation is security of property—we have shown too, that labour is the most sacred of properties. It results, therefore, that in any state of society in which the laws did not equally protect the capitalist and the labourer as free exchangers, each having the most absolute command over his property, compatible with a due regard to the rights of the other,—in such a state, where there was no real freedom and no real security, there would be very imperfect production; and production being imperfect, all men, the capitalists and the labourers, would be equally destitute, weak, ignorant, and miserable.

When the great body of the people of a country are so generally educated as to know that it is the interest of the humblest and the poorest that property shall be secure, there will be little occasion for fencing round property with guards, against the secret violence of the midnight robber, or the open daring of the noonday mob. " It is an enlightened moral public sentiment that must spread its wings over our dwellings, and plant a watchman at our doors."* A very little insecurity destroys the working of capital. The cloth trade of Verviers, a town in France, was utterly ruined, because the morals of the people in the town were so bad, and the police so ineffectual, that the thefts in the various stages of the manufacture amounted to eight per cent. upon the whole quantity produced. The trade of the place, therefore, was destroyed; and the capital went to encourage labour in places where the rights of property were better respected.

But, generally speaking, the security of property is not so much weakened by plunder, as by those incessant contentions which harass the march of capital and labour; and keep up an irritation between the classes of the capitalists and the labourers, who ought to be united in

* Everett's Address to the Working Man's Party.

the most intimate compact for a common good. These irritations most frequently exhibit themselves in the shape of combinations for the advance of wages. We have no hesitation in declaring our opinion that it is the positive duty of the working-man to obtain as high wages as he can extract out of the joint products of capital and labour ; and that he has an equal right to unite with other workmen in making as good a bargain as he can, consistently with the rights of others, for his contribution of industry to the business of production. But it is also necessary for us to declare our conviction that, in too many cases, the working men attempt an object which no single exertion, and no union however formidable or complete, can ever accomplish. They attempt to force wages beyond the point at which they could be maintained, with reference to the demand for the article produced ; - and if they succeed they extinguish the demand, and therefore extinguish the power of working at any wages. They drive the demand, and therefore the supply, into new channels ;—and they thrust out capital from amongst them, to work in other places where it can work with freedom and security. Above all, such combinations, and the resistance which they call up, have a tendency to loosen the bonds of mutual regard which ought to subsist between capitalists and labourers. Their real interests are one and the same.

All men are united in one bond of interests, and rights, and duties ; and although each of us have particular interests, the parts which we play in society are so frequently changing, that under one aspect we have each an interest contrary to that which we have under another aspect. It is in this way that we find ourselves suddenly bound closely with those against whom we thought ourselves opposed a moment before ; and thus no class can ever be said to be inimical to another class. In the midst, too, of all these instantaneous conflicts and unions, we are all interchangeably related in the double interest of capitalists and consumers,—that is, we have each and all an interest that property shall be respected, and that production shall be carried forward to its utmost point of

perfection, so as to make its products accessible to all.
The power of production, in its greatest developments
of industry, is really addressed to the satisfaction of the
commonest wants.   If production, as in despotic countries, were principally labouring that some men might
wear cloth of gold whilst others went naked, then we
should say that production was exclusively for the rich
oppressor.   But, thank God, the man who *exclusively*
wears " purple and fine linen every day" has ceased to
exist.   The looms do not work for him alone, but for
the great mass of the people.   It is to the staple articles
of consumption that the capitals of manufactures and
commerce address their employment.   Their employment depends upon the ability of the great body of the
people to purchase what they produce.   The courtiers
of the fifteenth century in France carried boxes of sugar-
plums in their pockets, which they offered to each other
as a constant compliment ; the courtiers of the next age
carried gingerbread in the same way ; and lastly, the
luxury of snuff drove out the sugar-plums and the gingerbread.   But the consumption of tobacco would never
have furnished employment to thousands, and a large
revenue to the state, if the use of snuff had rested with
the courtiers. The producers, consequently, having found
the largest, and therefore the most wealthy class of consumers amongst the working men, care little whether
the Peer wears a silk or a velvet coat, so that the Peasant
has a clean shirt.   When capital and labour work with
freedom and security, the wants of all are supplied, because there is cheap production.   It is a bad state of
society where

" One flaunts in rags, one flutters in brocade."

Those who like the brocade may still wear it in a state
of things where the rights of industry are understood ;
but the rags, taking the average condition of the members of society, are banished to the lands from which
capital is driven,—while those who labour with skill,
and therefore with capital, have decent clothes, comfortable dwellings, wholesome food, abundant fuel, medical

aid in sickness, the comfort and amusement of books in health. These goods, we have no hesitation in saying, all depend upon the security of property ; and he that would destroy that security by force or fraud is the real destroyer of the comforts of those humbler classes whose rights he pretends to advocate.

The principles which *we* maintain, that the interests of all men, and of the poorer classes especially, are necessarily advanced in a constantly increasing measure by the increase of capital and skill, have been put so strikingly by a philosophical writer, that we cannot forbear quoting so valuable an authority in support and illustration of our opinions :—

" The advantages conferred by the augmentation of our physical resources, through the medium of increased knowledge and improved art, have this peculiar and remarkable property,—that they are in their nature diffusive, and cannot be enjoyed in any exclusive manner by a few. An eastern despot may extort the riches and monopolize the art of his subjects for his own personal use ; he may spread around him an unnatural splendour and luxury, and stand in strange and preposterous contrast with the general penury and discomfort of his people ; he may glitter in jewels of gold and raiment of needle-work ; but the wonders of well contrived and executed manufacture which we use daily, and the comforts which have been invented, tried, and improved upon by thousands, in every form of domestic convenience, and for every ordinary purpose of life, can never be enjoyed by him. To produce a state of things in which the physical advantages of civilized life can exist in a high degree, the stimulus of increasing comforts and constantly elevated desires must have been felt by millions ; since it is not in the power of a few individuals to create that wide demand for useful and ingenious applications, which alone can lead to great and rapid improvements, unless backed by that arising from the speedy diffusion of the same advantages among the mass of mankind."*

* Herschel's Discourse on the Study of Natural Philosophy.

And, indeed, when we look at the operations of production as connected with the comforts of the working men of England, and generally with the comforts of what are called the productive classes, let it not be forgotten that the accumulations of property in the hands of these classes amount to an incomparably higher sum than the accumulations of those who receive the revenues of capital without employing it themselves. Who, then, should desire the protection of property?—who, in fact, do desire it? In 1830 there were 280,356 persons receiving half-yearly payments as fundholders. Of these, 85,767 were entitled to dividends not exceeding 5*l.* each, 45,147 to dividends above 5*l.* and not exceeding 10*l.*, 98,946 to dividends above 10*l.* and not exceeding 50*l.*, 26,205 to dividends above 50*l.* and not exceeding 100*l.*, 14,816 to dividends above 100*l.* and not exceeding 200*l.*, 4523 to dividends above 200*l.* and not exceeding 300*l.*, 2819 to dividends above 300*l.* and not exceeding 500*l.*, and 2133 to dividends exceeding 500*l.* Thus, 256,065 people who are public fundholders receive each an annual sum not exceeding 100*l.* a year from these funds, while not more than twenty-five thousand persons are entitled to dividends above that sum. By far the larger number of these fundholders, therefore, belong to the middle class.

But let us look a little farther as to the actual amount of accumulated capital, in the shape of *money*, in the hands of the working classes. The capital invested in 1842, in the Savings Banks of the United Kingdom, amounted to twenty-three millions six hundred and ninety-three thousand pounds. This capital was the property of eight hundred and seventy-four thousand seven hundred and fifteen depositors. Of this number only about one in fifteen held deposits above one hundred pounds. There are thus above a million (1,061,361) of people (including the fundholders under one hundred pounds, with the depositors in the Savings Banks) who cannot be pronounced *rich* in the common sense of the word, but who have what is commonly called " a stake in the country." But there are even additions still to

M

be made to the large number who have *monied* capital invested in public securities. The capital of Friendly Societies deposited in Savings Banks amounts to one million one hundred and twenty-one thousand pounds; there are besides nearly four hundred Friendly Societies whose investments amount to nearly one million five hundred thousand pounds, and who have a direct account with the National Debt Commissioners; and thus, probably, half a million people, members of Friendly Societies, in addition to the million of small fundholders and other Savings Banks depositors, have " a stake in the country." A stake in the country! Who has not a stake in the country? The humblest man who has shelter, and clothes, and food for a single day, has a stake in the country; because the *stakes*, the " plants," of other men ensure that he shall have food and clothes and shelter the next day;—that if misfortune happen, he shall be maintained;—and that, besides this indirect interest in the stakes of others, he may obtain by industry a positive stake himself—become a capitalist, and learn then, that labour and capital are *not* natural enemies, but the real partners in the production of riches and happiness.

THE END.

LONDON: WILLIAM CLOWES AND SONS, STAMFORD STREET.

THE EVOLUTION
OF CAPITALISM

Allen, Zachariah. **The Practical Tourist,** Or Sketches of the State of the Useful Arts, and of Society, Scenery, &c. &c. in Great-Britain, France and Holland. Providence, R.I., 1832. Two volumes in one.

Bridge, James Howard. **The Inside History of the Carnegie Steel Company:** A Romance of Millions. New York, 1903.

Brodrick, J[ames]. **The Economic Morals of the Jesuits:** An Answer to Dr. H. M. Robertson. London, 1934.

Burlamaqui, J[ean-] J[acques]. **The Principles of Natural and Politic Law.** Cambridge, Mass., 1807. Two volumes in one.

Capitalism and Fascism: Three Right-Wing Tracts, 1937-1941. New York, 1972.

Corey, Lewis. **The Decline of American Capitalism.** New York, 1934.

[Court, Pieter de la]. **The True Interest and Political Maxims, of the Republic of Holland.** Written by that Great Statesman and Patriot, John de Witt. To which is prefixed, (never before printed) Historical Memoirs of the Illustrious Brothers Cornelius and John de Witt, by John Campbell. London, 1746.

Dos Passos, John R. **Commercial Trusts:** The Growth and Rights of Aggregated Capital. An Argument Delivered Before the Industrial Commission at Washington, D.C., December 12, 1899. New York, 1901.

Fanfani, Amintore. **Catholicism, Protestantism and Capitalism.** London, 1935.

Gaskell, P[eter]. **The Manufacturing Population of England:** Its Moral, Social, and Physical Conditions, and the Changes Which Have Arisen From the Use of Steam Machinery; With an Examination of Infant Labour. London, 1833.

Göhre, Paul. **Three Months in a Workshop:** A Practical Study. London, 1895.

Greeley, Horace. **Essays Designed to Elucidate the Science of Political Economy,** While Serving to Explain and Defend the Policy of Protection to Home Industry, As a System of National Cooperation for the Elevation of Labor. Boston, 1870.

Grotius, Hugo. **The Freedom of the Seas,** Or, The Right Which Belongs to the Dutch to Take Part in the East Indian Trade. Translated with Revision of the Latin Text of 1633 by Ralph Van Deman Magoffin. New York, 1916.

Hadley, Arthur Twining. **Economics:** An Account of the Relations Between Private Property and Public Welfare. New York, 1896.

Knight, Charles. **Capital and Labour;** Including *The Results of Machinery.* London, 1845.

de Malynes, Gerrard. **Englands View, in the Unmasking of Two Paradoxes:** With a Replication unto the Answer of Maister John Bodine. London, 1603. New Introduction by Mark Silk.

Marquand, H. A. **The Dynamics of Industrial Combination.** London, 1931.

Mercantilist Views of Trade and Monopoly: Four Essays, 1645-1720. New York, 1972.

Morrison, C[harles]. **An Essay on the Relations Between Labour and Capital.** London, 1854.

Nicholson, J. Shield. **The Effects of Machinery on Wages.** London, 1892.

One Hundred Years' Progress of the United States: With an Appendix Entitled Marvels That Our Grandchildren Will See; or, One Hundred Years' Progress in the Future. By Eminent Literary Men, Who Have Made the Subjects on Which They Have Written Their Special Study. Hartford, Conn., 1870.

The Poetry of Industry: Two Literary Reactions to the Industrial Revolution, 1755/1757. New York, 1972.

Pre-Capitalist Economic Thought: Three Modern Interpretations. New York, 1972.

Promoting Prosperity: Two Eighteenth Century Tracts. New York, 1972.

Proudhon, P[ierre-] J[oseph]. **System of Economical Contradictions**: Or, The Philosophy of Misery. (Reprinted from *The Works of P. J. Proudhon*, Vol. IV, Part I.) Translated by Benj. R. Tucker. Boston, 1888.

Religious Attitudes Toward Usury: Two Early Polemics. New York, 1972.

Roscher, William. **Principles of Political Economy.** New York, 1878. Two volumes in one.

Scoville, Warren C. **Revolution in Glassmaking**: Entrepreneurship and Technological Change in the American Industry, 1880-1920. Cambridge, Mass., 1948.

Selden, John. **Of the Dominion, Or, Ownership of the Sea.** Written at First in Latin, and Entituled *Mare Clausum.* Translated by Marchamont Nedham. London, 1652.

Senior, Nassau W. **Industrial Efficiency and Social Economy.** Original Manuscript Arranged and Edited by S. Leon Levy. New York, 1928. Revised Preface by S. Leon Levy. Two volumes in one.

Spann, Othmar. **The History of Economics.** Translated from the 19th German Edition by Eden and Cedar Paul. New York, 1930.

The Usury Debate After Adam Smith: Two Nineteenth Century Essays. New York, 1972. New Introduction by Mark Silk.

The Usury Debate in the Seventeenth Century: Three Arguments. New York, 1972.

Varga, E[ugen]. **Twentieth Century Capitalism.** Translated from the Russian by George H. Hanna. Moscow, [1964].

Young, Arthur. **Arthur Young on Industry and Economics:** Being Excerpts from Arthur Young's Observations on the State of Manufactures and His Economic Opinions on Problems Related to Contemporary Industry in England. Arranged by Elizabeth Pinney Hunt. Bryn Mawr, Pa., 1926.

Db